Jozef Duda

The parametric optimization of a time delay system

Jozef Duda

The parametric optimization of a time delay system

The Lyapunov functionals for time delay systems

LAP LAMBERT Academic Publishing

Impressum / Imprint

Bibliografische Information der Deutschen Nationalbibliothek: Die Deutsche Nationalbibliothek verzeichnet diese Publikation in der Deutschen Nationalbibliografie; detaillierte bibliografische Daten sind im Internet über http://dnb.d-nb.de abrufbar.

Alle in diesem Buch genannten Marken und Produktnamen unterliegen warenzeichen-, marken- oder patentrechtlichem Schutz bzw. sind Warenzeichen oder eingetragene Warenzeichen der jeweiligen Inhaber. Die Wiedergabe von Marken, Produktnamen, Gebrauchsnamen, Handelsnamen, Warenbezeichnungen u.s.w. in diesem Werk berechtigt auch ohne besondere Kennzeichnung nicht zu der Annahme, dass solche Namen im Sinne der Warenzeichen- und Markenschutzgesetzgebung als frei zu betrachten wären und daher von jedermann benutzt werden dürften.

Bibliographic information published by the Deutsche Nationalbibliothek: The Deutsche Nationalbibliothek lists this publication in the Deutsche Nationalbibliografie; detailed bibliographic data are available in the Internet at http://dnb.d-nb.de.

Any brand names and product names mentioned in this book are subject to trademark, brand or patent protection and are trademarks or registered trademarks of their respective holders. The use of brand names, product names, common names, trade names, product descriptions etc. even without a particular marking in this works is in no way to be construed to mean that such names may be regarded as unrestricted in respect of trademark and brand protection legislation and could thus be used by anyone.

Coverbild / Cover image: www.ingimage.com

Verlag / Publisher:
LAP LAMBERT Academic Publishing
ist ein Imprint der / is a trademark of
OmniScriptum GmbH & Co. KG
Heinrich-Böcking-Str. 6-8, 66121 Saarbrücken, Deutschland / Germany
Email: info@lap-publishing.com

Herstellung: siehe letzte Seite /
Printed at: see last page
ISBN: 978-3-659-62252-6

The table of content

Introduction 3

1. A linear retarded type time delay system 4

1.1 A mathematical model of a linear retarded type time delay system 4

1.2 A Lyapunov functional 7

1.3 Determination of the Lyapunov functional 8

1.4 An example. An inertial system with a delay and a P controller 11

1.5 An example. An inertial system with a delay and an I controller 15

2. A linear system with both lumped and distributed retarded type time delay 24

2.1 A mathematical model of a linear system with both lumped and distributed retarded type time delay 24

2.2 Determination of the Lyapunov functional 25

2.3 An example 30

3. A linear neutral system with k non commensurate delays 35

3.1 A mathematical model of a linear neutral system with k-non-commensurate delays 35

3.2 A Lyapunov functional 38

3.3 Determination of the Lyapunov functional 40

3.3.1 The solution of equations (3.29) to (3.35) for k=1 41

3.3.2 The solution of equations (3.29) to (3.35) for k=2 43

3.4 An example. An inertial system with a delay and a PD controller 46

3.5 An example. A system with two delays and a PD controller 48

4. A linear system with a retarded type time-varying delay 51

4.1 A mathematical model of a linear system with a retarded type time-varying delay 51

4.2 A Lyapunov functional 52

4.3 Determination of the Lyapunov functional 53

4.4 The example. An inertial system with a delay and a P controller 58

5. A linear neutral system with a time-varying delay 61

5.1 A mathematical model of a linear neutral system with a time-varying delay 61

5.2 A Lyapunov functional 63

5.3 Determination of the Lyapunov functional 65

5.4 An example. An inertial system with a delay and a PD controller 70

Bibliography 74

Introduction

The Lyapunov functionals are used to test the stability of the systems. The Lyapunov functionals are also applied in calculation of the robustness bounds for uncertain time delay systems. To this end one uses the quadratic functionals, which are expressed by means of Lyapunov matrix, depended on the fundamental matrix of a time delay system. The descriptor model transformation and the decomposition technique are also used. The stability criteria are formulated in the form of linear matrix inequalities (LMIs). The Lyapunov functionals are also used in computation of the exponential estimates for the solutions of the time delay systems.

There are papers whose regard the quadratic Lyapunov functionals for the system with a time delay, such that their coefficients are given by the analytical formulas. To construct such Lyapunov functional, its time derivative on the trajectory of the system with a time delay is computed and equated with the negatively definite quadratic form of a system state. For the first time such Lyapunov functional, for the case of the retarded type time delay linear system with one delay, was introduced by Repin [11].

The Lyapunov quadratic functionals are also used to calculation of a value of a quadratic performance index of quality in the process of the parametric optimization for the time delay systems. The value of that functional at the initial state of the time delay system is equal to the value of a quadratic performance index of quality. To calculate the value of a performance index of quality one needs the formulas of the Lyapunov functional coefficients.

In this book the method proposed by Repin is applied to obtain the Lyapunov functionals, with coefficients given by the analytical formulas, to the system with retarded type time delay, to the system with both lumped and distributed delay, to the neutral systems with one and two delays, to the system with a retarded type time-varying delay and to the neutral system with a time-varying delay. The example of using of the Lyapunov functionals to calculation of the performance index value in the parametric optimization for linear systems with a time delay are given.

Chapter 1

A linear retarded type time delay system

1.1 A mathematical model of a linear retarded type time delay system

Let us consider a linear system with a retarded type time delay whose dynamics is described by a functional-differential equation (FDE)

$$\begin{cases} \frac{dx(t)}{dt} = Ax(t) + Bx(t-r) \\ x(t_0) = x_0 \\ x(t_0 + \theta) = \Phi(\theta) \end{cases} \tag{1.1}$$

$t \geq t_0$, $\theta \in [-r, 0)$, $r \geq 0$, $A, B \in \mathbb{R}^{n \times n}$, $x_0 \in \mathbb{R}^n$, $\Phi \in L^2([-r, 0), \mathbb{R}^n)$, where $L^2([-r, 0), \mathbb{R}^n)$ is a space of Lebesgue square integrable functions on interval $[-r, 0)$ with values in $\mathbb{R}^n$. The space of initial data is given by the Cartesian product $\mathbb{R}^n \times L^2([-r, 0), \mathbb{R}^n)$.

Definition 1. The function $x_t \in L^2([-r, 0), \mathbb{R}^n)$ is called a shifted restriction of x to an interval $[t-r, t)$ and is defined by a formula

$$x_t(\theta) := x(t + \theta) \tag{1.2}$$

for $t \geq t_0$, $\theta \in [-r, 0)$.

Definition 2. The norm in $L^2([-r, 0), \mathbb{R}^n)$ is defined by

$$\| \Phi \|_{L^2}^2 = \int_{-r}^{0} \left(\| \Phi(t) \|_{\mathbb{R}^n}^2 \right) dt \tag{1.3}$$

where $\| \cdot \|_{\mathbb{R}^n}$ is an arbitrary norm in $\mathbb{R}^n$.

Lemma 3. *There holds the relationship*

$$\frac{\partial x_t(\theta)}{\partial t} = \frac{\partial x_t(\theta)}{\partial \theta} \tag{1.4}$$

Proof.

$$x_t(\theta) = x(t + \theta) \quad for \, t \geq t_0, \, \theta \in [-r, 0)$$

4

$$\frac{\partial x_t(\theta)}{\partial t} = \frac{\partial x(t+\theta)}{\partial t} = \frac{\partial x(\xi)}{\partial \xi}\frac{\partial \xi}{\partial t} = \frac{\partial x(\xi)}{\partial \xi} \quad for\ \xi = t+\theta$$

$$\frac{\partial x_t(\theta)}{\partial \theta} = \frac{\partial x(t+\theta)}{\partial \theta} = \frac{\partial x(\xi)}{\partial \xi}\frac{\partial \xi}{\partial \theta} = \frac{\partial x(\xi)}{\partial \xi} \quad for\ \xi = t+\theta$$

hence

$$\frac{\partial x_t(\theta)}{\partial t} = \frac{\partial x_t(\theta)}{\partial \theta}$$

□

A solution of the functional-differential equation (1.1) with initial value (x_0, Φ) or simply a solution through (x_0, Φ) is an absolutely continuous function defined for $t \geq t_0 - r$ with values in $\mathbb{R}^n$.

$$x(\cdot; (x_0, \Phi)) \in W^{1,2}([t_0 - r, \infty), \mathbb{R}^n) \tag{1.5}$$

where $W^{1,2}([t_0 - r, \infty), \mathbb{R}^n)$ is a space of all absolutely continuous functions with derivatives in a space of Lebesgue square integrable functions on an interval $[t_0 - r, \infty)$ taking values in $\mathbb{R}^n$.

Definition 4. The function $x_t(\cdot; (x_0, \Phi)) \in W^{1,2}([-r, 0), \mathbb{R}^n)$ is called a shifted restriction of $x(\cdot; (x_0, \Phi))$ to an interval $[t - r, t)$ and is given by a formula

$$x_t(\theta; (x_0, \Phi)) := x(t + \theta; (x_0, \Phi)) \tag{1.6}$$

for $t \geq t_0$, $\theta \in [-r, 0)$

$$x_{t_0}(\cdot; (x_0, \Phi)) = \Phi \tag{1.7}$$

where $W^{1,2}([-r, 0), \mathbb{R}^n)$ is a space of all absolutely continuous functions with derivatives in a space of Lebesgue square integrable functions on interval $[-r, 0)$ with values in $\mathbb{R}^n$.

Definition 5. The norm in $W^{1,2}([-r, 0), \mathbb{R}^n)$ is defined by

$$\| \Phi \|_{W^{1,2}}^2 = \int_{-r}^{0} \left(\| \Phi(t) \|_{\mathbb{R}^n}^2 + \| \frac{d\Phi(t)}{dt} \|_{\mathbb{R}^n}^2 \right) dt \tag{1.8}$$

where $\| \cdot \|_{\mathbb{R}^n}$ is an arbitrary norm in $\mathbb{R}^n$.

One can obtain a solution of FDE (1.1) using a step method. The step method is a basic method for solving FDE with a lumped delay. A solution is found on successive intervals, one after another, by solving an ordinary equation without delay in each interval.

For $t \in [t_0, t_0 + r]$ the equation (1.1) takes a form

$$\begin{cases} \frac{dx(t)}{dt} = Ax(t) + B\Phi(t - r) \\ x(t_0) = x_0 \end{cases} \tag{1.9}$$

5

A solution of an equation (1.9) is given by a term

$$x(t) = e^{A(t-t_0)}x_0 + \int_{t_0}^{t} e^{A(t-\tau)}B\Phi(\tau - r)d\tau \qquad (1.10)$$

$$\Psi(t) = x(t) \qquad (1.11)$$

$$x(t_0 + r) = x_1 \qquad (1.12)$$

For $t \in [t_0 + r, t_0 + 2r]$ the equation (1.1) takes a form

$$\begin{cases} \frac{dx(t)}{dt} = Ax(t) + B\Psi(t-r) \\ x(t_0 + r) = x_1 \end{cases} \qquad (1.13)$$

and so on. By this procedure one can construct the solution in any finite interval.

Using the formula (1.2) one can write the equation (1.1) in the form

$$\begin{cases} \frac{dx(t)}{dt} = Ax(t) + Bx_t(-r) \\ x(t_0) = x_0 \\ x_{t_0} = \Phi \in L^2([-r,0), \mathbb{R}^n) \end{cases} \qquad (1.14)$$

for $t \geq t_0$.

The norm of an initial value (x_0, Φ) is given by

$$\| (x_0, \Phi) \| = \sqrt{\| x_0 \|_{\mathbb{R}^n}^2 + \| \Phi \|_{L^2}^2} \qquad (1.15)$$

The theorems of existence, continuous dependence and uniqueness of solutions of the equation (1.14) are given in [9].

Definition 6. The zero solution of (1.14) is **stable** if for any $\varepsilon > 0$ there is a $\delta > 0$ such that

$$\| (x_0, \Phi) \| < \delta$$

implies $\| x_t(\cdot; (x_0, \Phi)) \| < \varepsilon$ for $t \geq t_0$.

Definition 7. The zero solution of (1.14) is **asymptotically stable** if

$$\| x_t(\cdot; (x_0, \Phi)) \| \to 0$$

as $t \to \infty$.

Definition 8. The zero solution of (1.14) is **exponentially stable** if there exists an $\eta > 0$ and an positive constant M such that

$$\| x_t(\cdot; (x_0, \Phi)) \| \leq M e^{-\eta t} \| (x_0, \Phi) \| \qquad (1.16)$$

for $t \geq t_0$.

The state of a system (1.14) is a vector

$$S(t) = \begin{bmatrix} x(t) \\ x_t \end{bmatrix} \tag{1.17}$$

for $t \geq t_0$ where $x_t \in L^2([-r,0),\mathbb{R}^n)$

The state space is defined by a formula

$$X = \mathbb{R}^n \times L^2([-r,0),\mathbb{R}^n) \tag{1.18}$$

$S = 0$ is an equilibrium point of the system (1.1).

In a parametric optimization problem is used an integral quadratic performance index of quality

$$J = \int_{t_0}^{\infty} x^T(t)x(t)dt \tag{1.19}$$

1.2 A Lyapunov functional

Definition 9. A functional $V : X \to \mathbb{R}$ is **positive definite** if and only if it is continuous and $V(x) > 0$ for $x \neq 0$ and $V(0) = 0$.

A functional $V : X \to \mathbb{R}$ is **negative definite** if and only if it is continuous and $V(x) < 0$ for $x \neq 0$ and $V(0) = 0$.

Definition 10. The time derivative of the functional $V(x(t),x_t)$ at $(x(t_0),\Phi)$ on a trajectory of a system (1.14) is defined by a formula

$$\frac{dV(x(t_0),\Phi)}{dt} = \limsup_{h \to 0} \frac{1}{h} \left[V\left(x(t_0+h),x_{t_0+h}\right) - V\left(x(t_0),\Phi\right) \right] \tag{1.20}$$

Definition 11. The functional $V : X \to \mathbb{R}$ is called the **Lyapunov functional** if

1. V is positive definite

2. V is differentiable

3. A time derivative of V computed according to a formula (1.20) on the trajectory of the system (1.14) is negative definite

Existence of the Lyapunov functional for the system (1.14) is a sufficient condition for asymptotic stability of its zero solution.

When the system (1.14) is asymptotically stable

$$\int_{t_0}^{\infty} \frac{dV(x(t),x_t)}{dt}dt = \lim_{t \to \infty} V(x(t),x_t) - \lim_{t \to t_0} V(x(t),x_t) =$$

$$= V\left(\lim_{t\to\infty}(x(t),x_t)\right) - V\left(\lim_{t\to t_0}(x(t),x_t)\right) =$$

$$= V(0) - V(x(t_0),\Phi) = -V(x(t_0),\Phi) \tag{1.21}$$

Assume that the time derivative of the Lyapunov functional V is given as a quadratic form

$$\frac{dV(x(t),x_t)}{dt} \equiv -x^T(t)x(t) \; for \, t \geq t_0 \tag{1.22}$$

It follows from (1.21) and (1.22) that

$$J = \int_{t_0}^{\infty} x^T(t)x(t)dt = V(x_0,\Phi) \tag{1.23}$$

Corollary 12. *If one constructs a Lyapunov functional such that its time derivative computed on the trajectory of the system (1.14) is given as a quadratic form (1.22) one can not only investigate the system (1.14) stability but also calculate a value of a square indicator of quality (1.23) of the parametric optimization problem.*

To calculate the value of the performance index (1.23), which is equal to the value of the Lyapunov functional at the initial state of the system (1.14), one needs a mathematical formula of that functional.

1.3 Determination of the Lyapunov functional

Let us consider a quadratic functional on X, given by a formula

$$V(x(t),x_t) = x^T(t)\alpha x(t) + \int_{-r}^{0} x^T(t)\beta(\theta)x_t(\theta)d\theta +$$

$$+ \int_{-r}^{0}\int_{\theta}^{0} x_t^T(\theta)\delta(\theta,\sigma)x_t(\sigma)d\sigma d\theta \tag{1.24}$$

for $t \geq t_0$, where $\alpha \in \mathbb{R}^{n \times n}$, $\beta \in C^1([-r,0],\mathbb{R}^{n \times n})$, $\delta \in C^1(\Omega,\mathbb{R}^{n \times n})$
$\Omega = \{(\theta,\sigma): \theta \in [-r,0], \sigma \in [\theta,0]\}$ C^1 is a space of continuous functions with continuous derivative.

Conjecture 13. *It is given a procedure of determination of the functional (1.24) coefficients to obtain the Lyapunov functional.*

The time derivative of the functional (1.24) on the trajectory of the system (1.14) is computed. This time derivative is defined by the formula (1.20). It is taken the following procedure. One computes the time derivative of each term of the right-hand-side of the formula (1.24) and one substitutes in place of $\frac{dx(t)}{dt}$ and $\frac{\partial x_t(\theta)}{\partial t}$ the following terms

$$\frac{dx(t)}{dt} = Ax(t) + Bx_t(-r) \tag{1.25}$$

$$\frac{\partial x_t(\theta)}{\partial t} = \frac{\partial x_t(\theta)}{\partial \theta} \tag{1.26}$$

In such a manner one attains

$$\frac{dV(x(t), x_t)}{dt} = x^T(t) \left[A^T \alpha + \alpha A + \frac{\beta(0) + \beta^T(0)}{2} \right] x(t) +$$

$$+ x^T(t) \left[2\alpha B - \beta(-r) \right] x_t(-r) +$$

$$+ \int_{-r}^{0} x^T(t) \left[A^T \beta(\theta) - \frac{d\beta(\theta)}{d\theta} + \delta^T(\theta, 0) \right] x_t(\theta) d\theta +$$

$$+ \int_{-r}^{0} x_t^T(-r) \left[B^T \beta(\theta) - \delta(-r, \theta) \right] x(t+\theta) d\theta +$$

$$- \int_{-r}^{0} \int_{\theta}^{0} x_t^T(\theta) \left[\frac{\partial \delta(\theta, \sigma)}{\partial \theta} + \frac{\partial \delta(\theta, \sigma)}{\partial \sigma} \right] x_t(\sigma) d\sigma d\theta \tag{1.27}$$

From equations (1.27) and (1.22) one obtains a system of equations

$$A^T \alpha + \alpha A + \frac{\beta(0) + \beta^T(0)}{2} = -I \tag{1.28}$$

$$2\alpha B - \beta(-r) = 0 \tag{1.29}$$

$$A^T \beta(\theta) - \frac{d\beta(\theta)}{d\theta} + \delta^T(\theta, 0) = 0 \tag{1.30}$$

$$B^T \beta(\theta) - \delta(-r, \theta) = 0 \tag{1.31}$$

$$\frac{\partial \delta(\theta, \sigma)}{\partial \theta} + \frac{\partial \delta(\theta, \sigma)}{\partial \sigma} = 0 \tag{1.32}$$

for $\theta \in [-r, 0]$, $\sigma \in [-r, 0]$.

A solution of a differential equation (1.32) is given in a form

$$\delta(\theta, \sigma) = \varphi(\theta - \sigma) \tag{1.33}$$

where $\varphi \in C^1([-r, r], \mathbb{R}^{n \times n})$.

9

From the equations (1.33) and (1.31) one obtains

$$\delta(-r, \theta) = \varphi(-r - \theta) = B^T \beta(\theta) \tag{1.34}$$

$$\varphi(\theta) = B^T \beta(-r - \theta) \tag{1.35}$$

$$\delta^T(\theta, 0) = \varphi^T(\theta) = \beta^T(-r - \theta)B \tag{1.36}$$

After putting (1.36) into (1.30) one attains a formula

$$\frac{d\beta(\theta)}{d\theta} = A^T \beta(\theta) + \beta^T(-r - \theta)B$$

for $\theta \in [-r, 0]$.

The derivative of the function $\beta(-\theta - r)$ with respect to θ is calculated

$$\frac{d\beta(-r - \theta)}{d\theta} = \frac{d\beta(\xi)}{d\xi} \frac{d\xi}{d\theta} = -\frac{d\beta(\xi)}{d\xi} =$$

$$= -A^T \beta(\xi) - \beta^T(-r - \xi)B = -A^T \beta(-r - \theta) - \beta^T(\theta)B \tag{1.37}$$

where

$$\xi = -r - \theta \tag{1.38}$$

A set of the differential equations is obtained

$$\begin{cases} \frac{d\beta(\theta)}{d\theta} = A^T \beta(\theta) + \beta^T(-r - \theta)B \\ \frac{d\beta(-r-\theta)}{d\theta} = -A^T \beta(-r - \theta) - \beta^T(\theta)B \end{cases} \tag{1.39}$$

for $\theta \in [-r, 0]$.

A new function is introduced

$$\kappa(\theta) = \beta(-\theta - r) \tag{1.40}$$

for $\theta \in [-r, 0]$.

The set of the differential equations (1.39) takes a form

$$\begin{cases} \frac{d\beta(\theta)}{d\theta} = A^T \beta(\theta) + \kappa^T(\theta)B \\ \frac{d\kappa(\theta)}{d\theta} = -A^T \kappa(\theta) - \beta^T(\theta)B \end{cases} \tag{1.41}$$

for $\theta \in [-r, 0]$.

The solution of the differential equations (1.41) satisfies a condition

10

$$\beta(\theta)|_{\theta=-\frac{r}{2}} = \kappa(\theta)|_{\theta=-\frac{r}{2}} \tag{1.42}$$

A formula (1.42) was obtained from (1.40).

From the equation (1.40) it results that

$$\kappa(-r) = \beta(0) \tag{1.43}$$

The formula (1.28) takes a form

$$A^T\alpha + \alpha A + \frac{\kappa(-r) + \kappa^T(-r)}{2} = -I \tag{1.44}$$

One obtains a set of algebraic equations

$$\begin{cases} A^T\alpha + \alpha A + \frac{\kappa(-r)+\kappa^T(-r)}{2} = -I \\ 2B^T\alpha - \beta^T(-r) = 0 \\ \beta(\theta)|_{\theta=-\frac{r}{2}} = \kappa(\theta)|_{\theta=-\frac{r}{2}} \end{cases} \tag{1.45}$$

The set of algebraic equations (1.45) allows for determination of the matrix α and the initial conditions of the system of differential equations (1.41).

From the equations (1.35) and (1.40) one attains

$$\varphi(\theta) = B^T\beta(-r-\theta) = B^T\kappa(\theta) \tag{1.46}$$

for $\theta \in [-r,0]$.

Taking into account (1.33) and (1.46) one obtains

$$\delta(\theta,\sigma) = B^T\kappa(\theta-\sigma) \tag{1.47}$$

In this way one obtained all coefficients of the functional (1.24). This coefficients depend on the matrices A and B of the system (1.1). The time derivative of the functional (1.24) is negative definite. When the matrices α, $\beta(\theta)$ and $\delta(\theta,\sigma)$ for $t \geq t_0, \theta \in [-r,0], \sigma \in [\theta,0]$ are positive definite then the functional (1.24) becomes the Lyapunov functional.

1.4 An example. An inertial system with a delay and a P controller.

Let us consider a first order inertial system with a delay described by an equation

$$\begin{cases} \frac{dx(t)}{dt} = -\frac{q}{T}x(t) + \frac{k_0}{T}u(t-r) \\ x(0) = x_o \\ x(\theta) = 0 \\ u(t) = -kx(t) + z_0 1(t) \end{cases} \tag{1.48}$$

11

$t \geq 0$, $x(t) \in \mathbb{R}$, $\theta \in [-r, 0)$, k, k_0, T, q, x_0, $z_0 \in \mathbb{R}$, $r \geq 0$. The parameter k_0 is a gain of a plant, k is a gain of a P controller, T is a system time constant, x_0 is an initial state of a system, z_0 is an amplitude of a disturbance. In the case $q = 1$ an equation (1.48) describes a static object and in the case $q = 0$ an equation (1.48) describes an astatic object.

One can reshape an equation (1.48) to a form

$$\begin{cases} \frac{dx(t)}{dt} = -\frac{q}{T}x(t) - \frac{k_0 k}{T}x(t - r) + \frac{k_0}{T}z_0 1(t - r) \\ x(0) = x_o \\ x(\theta) = 0 \end{cases} \tag{1.49}$$

for $t \geq 0$.

An equilibrium point of the system (1.49) is given by a term

$$x^* = \frac{k_0 z_0}{k_0 k + q} \tag{1.50}$$

One searches for a parameter k whose minimize an integral quadratic performance index

$$J = \int_0^\infty \left(x(t) - \frac{k_0 z_0}{k_0 k + q} \right)^2 dt \tag{1.51}$$

The time axes is divided into two intervals $[0, r)$ and $[r, \infty)$ and one calculates a performance index value (1.51) for each interval.

$$J = J_1 + J_2 \tag{1.52}$$

$$J_1 = \int_0^r \left(x(t) - \frac{k_0 z_0}{k_0 k + q} \right)^2 dt \tag{1.53}$$

$$J_2 = \int_r^\infty \left(x(t) - \frac{k_0 z_0}{k_0 k + q} \right)^2 dt \tag{1.54}$$

1. For $t \in [0, r)$ the equation (1.49) takes a form

$$\begin{cases} \frac{dx(t)}{dt} = -\frac{q}{T}x(t) \\ x(0) = x_o \end{cases} \tag{1.55}$$

A solution of an equation (1.55) is given by a term

$$x(t) = x_0 e^{-\frac{q}{T}t} \tag{1.56}$$

for $t \in [0, r)$.

A value of the performance index J_1 is given by a formula

$$J_1 = \left(\frac{k_0 z_0}{k_0 k + q}\right)^2 r + \frac{2k_0 T z_0 x_0}{q(k_0 k + q)}\left(e^{-\frac{qr}{T}} - 1\right) + \frac{T x_0^2}{2q}\left(1 - e^{-\frac{2qr}{T}}\right) \tag{1.57}$$

2. For $t \in [r, \infty)$ the equation (1.49) takes a form

$$\begin{cases} \frac{dx(t)}{dt} = -\frac{q}{T}x(t) - \frac{k_0 k}{T}x(t - r) + \frac{k_0 z_0}{T} \\ x(r) = x_0 e^{-\frac{qr}{T}} \\ x(r + \theta) = x_0 e^{-\frac{q(r+\theta)}{T}} \end{cases} \tag{1.58}$$

for $\theta \in [-r, 0)$.

A new variable is introduced

$$y(t) = x(t) - \frac{k_0 z_0}{k_0 k + q} \tag{1.59}$$

Taking a term (1.59) into account a set of the equations (1.58) takes a form

$$\begin{cases} \frac{dy(t)}{dt} = -\frac{q}{T}y(t) - \frac{k_0 k}{T}y(t - r) \\ y(r) = x_0 e^{-\frac{qr}{T}} - \frac{k_0 z_0}{k_0 k + q} \\ y(r + \theta) = x_0 e^{-\frac{q(r+\theta)}{T}} - \frac{k_0 z_0}{k_0 k + q} \end{cases} \tag{1.60}$$

for $t \geq r$ and $\theta \in [-r, 0)$.

A performance index J_2 is given by a formula

$$J_2 = \int_r^\infty y^2(t)dt \tag{1.61}$$

According to a term (1.54) a performance index value is given by a formula

$$J_2 = \int_r^\infty y^2(t)dt = V(S(r)) \tag{1.62}$$

A Lyapunov functional V is defined by a formula

$$V(y(t), y(t + \theta)) = \alpha y^2(t) + \int_{-r}^0 y(t)\beta(\theta)y(t + \theta)d\theta + \int_{-r}^0\int_{-r}^0 y(t + \theta)\delta(\theta, \sigma)y(t + \sigma)d\sigma d\theta$$

A set of a differential equation (1.41) takes a form

$$\begin{bmatrix} \frac{d\beta(\theta)}{d\theta} \\ \frac{d\kappa(\theta)}{d\theta} \end{bmatrix} = \begin{bmatrix} -\frac{q}{T} & -\frac{k_0 k}{T} \\ \frac{k_0 k}{T} & \frac{q}{T} \end{bmatrix}\begin{bmatrix} \beta(\theta) \\ \kappa(\theta) \end{bmatrix} \tag{1.63}$$

A fundamental matrix of a system (1.63) takes a form

$$R(\theta) = \begin{bmatrix} \cosh\lambda\theta - \frac{q}{T\lambda}\sinh\lambda\theta & -\frac{k_0k}{T\lambda}\sinh\lambda\theta \\ \frac{k_0k}{T\lambda}\sinh\lambda\theta & \cosh\lambda\theta + \frac{q}{T\lambda}\sinh\lambda\theta \end{bmatrix} \qquad (1.64)$$

where

$$\lambda = \frac{\sqrt{q^2 - k_0^2 k^2}}{T} \qquad (1.65)$$

The set of the algebraic equations (1.45) takes a form

$$\begin{cases} -2\frac{q}{T}\alpha + \kappa(-r) = -1 \\ 2\alpha\frac{k_0k}{T} + \beta(-r) = 0 \\ \left[\cosh\frac{\lambda r}{2} - \frac{q+k_0k}{T\lambda}\sinh\frac{\lambda r}{2}\right]\beta(-r) + \left[-\cosh\frac{\lambda r}{2} - \frac{q+k_0k}{T\lambda}\sinh\frac{\lambda r}{2}\right]\kappa(-r) = 0 \end{cases} \qquad (1.66)$$

From an equation (1.66) one obtains a parameter α and the initial conditions of the differential equation(1.63).

$$\alpha = \frac{\cosh\frac{\lambda r}{2} + \frac{q+k_0k}{T\lambda}\sinh\frac{\lambda r}{2}}{2\left(\lambda\sinh\frac{\lambda r}{2} + \frac{q+k_0k}{T}\cosh\frac{\lambda r}{2}\right)} \qquad (1.67)$$

$$\beta(-r) = \frac{\frac{k_0k}{T}\left(\cosh\frac{\lambda r}{2} + \frac{q+k_0k}{T\lambda}\sinh\frac{\lambda r}{2}\right)}{\lambda\sinh\frac{\lambda r}{2} + \frac{q+k_0k}{T}\cosh\frac{\lambda r}{2}} \qquad (1.68)$$

$$\kappa(-r) = \frac{-\frac{k_0k}{T}\left(\cosh\frac{\lambda r}{2} - \frac{q+k_0k}{T\lambda}\sinh\frac{\lambda r}{2}\right)}{\lambda\sinh\frac{\lambda r}{2} + \frac{q+k_0k}{T}\cosh\frac{\lambda r}{2}} \qquad (1.69)$$

Having a fundamental matrix (1.64) and the initial conditions of the differential equation(1.63) one obtains

$$\beta(\theta) = \frac{k_0k}{T\left(\lambda\sinh\frac{\lambda r}{2} + \frac{q+k_0k}{T}\cosh\frac{\lambda r}{2}\right)}\left[\left(\frac{q+k_0k}{T\lambda}\cosh\frac{\lambda r}{2} - \sinh\frac{\lambda r}{2}\right)\sinh\lambda\theta +\right.$$

$$\left. +\left(\frac{q+k_0k}{T\lambda}\sinh\frac{\lambda r}{2} - \cosh\frac{\lambda r}{2}\right)\cosh\lambda\theta\right] \qquad (1.70)$$

$$\kappa(\theta) = -\frac{k_0k}{T\lambda}\sinh\lambda\theta +$$

$$-\frac{k_0k}{T\left(\lambda\sinh\frac{\lambda r}{2} + \frac{q+k_0k}{T}\cosh\frac{\lambda r}{2}\right)}\left(\cosh\frac{\lambda r}{2} + \frac{q+k_0k}{T\lambda}\sinh\frac{\lambda r}{2}\right)\cosh\lambda\theta \qquad (1.71)$$

$$\delta(\theta,\sigma) = \frac{k_0^2 k^2}{T^2\lambda}\sinh\lambda(\theta - \sigma) +$$

$$+\frac{k_0^2 k^2}{T^2\left(\lambda \sinh\frac{\lambda r}{2}+\frac{q+k_0 k}{T}\cosh\frac{\lambda r}{2}\right)}\left(\cosh\frac{\lambda r}{2}+\frac{q+k_0 k}{T\lambda}\sinh\frac{\lambda r}{2}\right)\cosh\lambda(\theta-\sigma) \tag{1.72}$$

Now a performance index value is calculated

$$J=\frac{k_0^2 z_0^2}{(q+k_0 k)^2\left(\lambda \sinh\frac{\lambda r}{2}+\frac{q+k_0 k}{T}\cosh\frac{\lambda r}{2}\right)}[\frac{q^2 r}{T^2\lambda}\sinh\frac{\lambda r}{2}+$$

$$+\frac{k_0^2 k^2+4qk_0 k-q^2}{2T\lambda\,(k_0 k-q)}\sinh\frac{\lambda r}{2}+\frac{q^2 r}{T\,(q-k_0 k)}\cosh\frac{\lambda r}{2}+\frac{1}{2}\cosh\frac{\lambda r}{2}]+$$

$$+\frac{k_0 z_0 x_0}{(q+k_0 k)\left(\lambda \sinh\frac{\lambda r}{2}+\frac{q+k_0 k}{T}\cosh\frac{\lambda r}{2}\right)}\left[-\frac{3q+k_0 k}{T\lambda}\sinh\frac{\lambda r}{2}-\cosh\frac{\lambda r}{2}\right]+$$

$$+\frac{x_0^2}{2\left(\lambda \sinh\frac{\lambda r}{2}+\frac{q+k_0 k}{T}\cosh\frac{\lambda r}{2}\right)}\left[\cosh\frac{\lambda r}{2}+\frac{q+k_0 k}{T\lambda}\sinh\frac{\lambda r}{2}\right] \tag{1.73}$$

1.5 An example. An inertial system with a delay and an I controller.

Let us consider a first order inertial system with a delay described by an equation

$$\begin{cases} \frac{dx(t)}{dt}=-\frac{1}{T}x(t)+\frac{k_0}{T}u(t-r) \\ x(0)=x_o \\ x(\theta)=0 \\ u(t)=-\frac{1}{T_i}\int_0^t x(\xi)d\xi+z_0 \end{cases} \tag{1.74}$$

$t\ge 0$, $x(t)\in\mathbb{R}$, $\theta\in[-r,0)$, T_i, k_0, T, x_0, $z_0\in\mathbb{R}$, $r\ge 0$. The parameter k_0 is a gain of a plant, T_i is a time of isodrome of an I controller, T is a system time constant, x_0 is an initial state of a system, z_0 is an amplitude of a disturbance.

One introduces the state variables $x_1(t)$ and $x_2(t)$ as follows

$$\begin{cases} x_1(t)=x(t) \\ x_2(t)=\frac{1}{T_i}\int_0^t x(\xi)d\xi \end{cases} \tag{1.75}$$

The set of equations (1.74) takes a form

$$\begin{cases} \frac{dx_1(t)}{dt} = -\frac{1}{T}x_1(t) + \frac{k_0}{T}u(t-r) \\ \frac{dx_2(t)}{dt} = \frac{1}{T_i}x_1(t) \\ x_1(0) = x_o \\ x_2(0) = 0 \\ x_1(\theta) = 0 \\ x_2(\theta) = 0 \\ u(t) = -x_2(t) + z_0 \end{cases} \tag{1.76}$$

for $t \geq 0$, $\theta \in [-r, 0)$.

One can reshape an equation (1.76) to a form

$$\begin{cases} \frac{dx_1(t)}{dt} = -\frac{1}{T}x_1(t) - \frac{k_0}{T}x_2(t-r) + \frac{k_0 z_0}{T} \\ \frac{dx_2(t)}{dt} = \frac{1}{T_i}x_1(t) \\ x_1(0) = x_o \\ x_2(0) = 0 \\ x_1(\theta) = 0 \\ x_2(\theta) = 0 \end{cases} \tag{1.77}$$

for $t \geq 0$, $\theta \in [-r, 0)$.

An equilibrium point of the system (1.77) is given by a term

$$\begin{cases} x_1^* = 0 \\ x_2^* = z_0 \end{cases} \tag{1.78}$$

One introduces a new variable

$$\begin{cases} y_1(t) = x_1(t) \\ y_2(t) = x_2(t) - z_0 \end{cases} \tag{1.79}$$

Taking a term (1.79) into account a set of the equations (1.77) takes a form

$$\begin{cases} \frac{dy_1(t)}{dt} = -\frac{1}{T}y_1(t) - \frac{k_0}{T}y_2(t-r) \\ \frac{dy_2(t)}{dt} = \frac{1}{T_i}y_1(t) \\ y_1(0) = x_o \\ y_2(0) = -z_0 \\ y_1(\theta) = 0 \\ y_2(\theta) = -z_0 \end{cases} \tag{1.80}$$

The equations (1.80) in a matrix form take a form

$$
\begin{cases}
\frac{dy(t)}{dt} = Ay(t) + By(t-r) \\[2mm]
y(0) = \begin{bmatrix} x_0 \\ -z_0 \end{bmatrix} \\[4mm]
y(\theta) = \begin{bmatrix} 0 \\ -z_0 \end{bmatrix}
\end{cases}
\tag{1.81}
$$

where

$$
A = \begin{bmatrix} -\frac{1}{T} & 0 \\ \frac{1}{T_i} & 0 \end{bmatrix}
\tag{1.82}
$$

$$
B = \begin{bmatrix} 0 & -\frac{k_0}{T} \\ 0 & 0 \end{bmatrix}
\tag{1.83}
$$

One searches for a parameter T_i whose minimize an integral quadratic performance index

$$
J = \int_0^\infty y^T(t)y(t)dt
\tag{1.84}
$$

The Lyapunov functional is given

$$
V(y(t), y(t+\theta)) = y^T(t)\alpha y(t) + \int_{-r}^0 y^T(t)\beta(\theta)y(t+\theta)d\theta +
$$

$$
+ \int_{-r}^0 \int_\theta^0 y^T(t+\theta)\delta(\theta,\sigma)y(t+\sigma)d\sigma d\theta
\tag{1.85}
$$

where

$$
\alpha = \begin{bmatrix} \alpha_{11} & \alpha_{12} \\ \alpha_{12} & \alpha_{22} \end{bmatrix}
\tag{1.86}
$$

$$
\beta(\theta) = \begin{bmatrix} \beta_{11}(\theta) & \beta_{12}(\theta) \\ \beta_{12}(\theta) & \beta_{22}(\theta) \end{bmatrix}
\tag{1.87}
$$

$$
\delta(\theta,\sigma) = \begin{bmatrix} \delta_{11}(\theta,\sigma) & \delta_{12}(\theta,\sigma) \\ \delta_{21}(\theta,\sigma) & \delta_{22}(\theta,\sigma) \end{bmatrix}
\tag{1.88}
$$

$$
J = \int_0^\infty y^T(t)y(t)dt = V(y(0), y(\theta))
\tag{1.89}
$$

The set of differential equations (1.41) takes a form

17

$$\begin{cases} \frac{d\beta_{11}(\theta)}{d\theta} = -\frac{1}{T}\beta_{11}(\theta) + \frac{1}{T_i}\beta_{21}(\theta) \\ \frac{d\beta_{21}(\theta)}{d\theta} = 0 \\ \frac{d\beta_{12}(\theta)}{d\theta} = -\frac{1}{T}\beta_{12}(\theta) + \frac{1}{T_i}\beta_{22}(\theta) - \frac{k_0}{T}\kappa_{11}(\theta) \\ \frac{d\beta_{22}(\theta)}{d\theta} = -\frac{k_0}{T}\kappa_{12}(\theta) \\ \frac{d\kappa_{11}(\theta)}{d\theta} = \frac{1}{T}\kappa_{11}(\theta) - \frac{1}{T_i}\kappa_{21}(\theta) \\ \frac{d\kappa_{21}(\theta)}{d\theta} = 0 \\ \frac{d\kappa_{12}(\theta)}{d\theta} = \frac{1}{T}\kappa_{12}(\theta) - \frac{1}{T_i}\kappa_{22}(\theta) + \frac{k_0}{T}\beta_{11}(\theta) \\ \frac{d\kappa_{22}(\theta)}{d\theta} = \frac{k_0}{T}\beta_{12}(\theta) \end{cases} \tag{1.90}$$

for $\theta \in [-r, 0]$, where

$$\kappa(\theta) = \beta(-\theta - r) \tag{1.91}$$

for $\theta \in [-r, 0]$

The two first equations of (1.45) takes a form

$$\begin{cases} -\frac{2}{T}\alpha_{11} + \frac{2}{T_i}\alpha_{12} + \kappa_{11}(-r) = -1 \\ -\frac{2}{T}\alpha_{12} + \frac{2}{T_i}\alpha_{22} + \kappa_{12}(-r) + \kappa_{21}(-r) = 0 \\ \kappa_{22}(-r) = -1 \\ \beta_{11}(-r) = 0 \\ \beta_{21}(-r) = 0 \\ -\frac{2k_0}{T}\alpha_{11} - \beta_{12}(-r) = 0 \\ -\frac{2k_0}{T}\alpha_{12} - \beta_{22}(-r) = 0 \end{cases} \tag{1.92}$$

Equations (1.90), (1.91) and (1.91) implies $\beta_{21}(\theta) = 0$, $\kappa_{21}(\theta) = 0$, $\beta_{11}(\theta) = 0$, $\kappa_{11}(\theta) = 0$ for $\theta \in [-r, 0]$.

The formula (1.87) takes a form

$$\beta(\theta) = \begin{bmatrix} 0 & \beta_{12}(\theta) \\ 0 & \beta_{22}(\theta) \end{bmatrix} \tag{1.93}$$

and

$$\kappa(\theta) = \begin{bmatrix} 0 & \kappa_{12}(\theta) \\ 0 & \kappa_{22}(\theta) \end{bmatrix} \tag{1.94}$$

The set of equations (1.90) takes a form

$$\begin{cases} \frac{d\beta_{12}(\theta)}{d\theta} = -\frac{1}{T}\beta_{12}(\theta) + \frac{1}{T_i}\beta_{22}(\theta) \\ \frac{d\beta_{22}(\theta)}{d\theta} = -\frac{k_0}{T}\kappa_{12}(\theta) \\ \frac{d\kappa_{12}(\theta)}{d\theta} = \frac{1}{T}\kappa_{12}(\theta) - \frac{1}{T_i}\kappa_{22}(\theta) \\ \frac{d\kappa_{22}(\theta)}{d\theta} = \frac{k_0}{T}\beta_{12}(\theta) \end{cases} \tag{1.95}$$

The fundamental matrix of solutions of equation (1.88) is given by

$$R(\theta) = \frac{1}{s_1^2 + s_2^2} \begin{bmatrix} r_{11}(\theta) & r_{12}(\theta) & r_{13}(\theta) & r_{14}(\theta) \\ r_{21}(\theta) & r_{22}(\theta) & r_{23}(\theta) & r_{24}(\theta) \\ r_{31}(\theta) & r_{32}(\theta) & r_{33}(\theta) & r_{34}(\theta) \\ r_{41}(\theta) & r_{42}(\theta) & r_{43}(\theta) & r_{44}(\theta) \end{bmatrix} \tag{1.96}$$

where

$$s_i = \frac{1}{T}\sqrt{\frac{\sqrt{1 + \frac{4k_0^2 T^2}{T_i^2}} + (-1)^i}{2}} \quad for \quad i = 1, 2 \tag{1.97}$$

$$r_{11}(\theta) = s_1^2 \cos s_1\theta - \frac{s_1}{T}\sin s_1\theta + s_2^2 \cosh s_2\theta - \frac{s_2}{T}\sinh s_2\theta \tag{1.98}$$

$$r_{21}(\theta) = \frac{k_0^2}{T^2 T_i}(-\frac{1}{s_1}\sin s_1\theta + \frac{1}{s_2}\sinh s_2\theta) \tag{1.99}$$

$$r_{31}(\theta) = \frac{k_0}{T T_i}(\cos s_1\theta - \cosh s_2\theta) \tag{1.100}$$

$$r_{41}(\theta) = \frac{k_0}{T}(s_1 \sin s_1\theta + \frac{1}{T}\cos s_1\theta + s_2 \sinh s_2\theta - \frac{1}{T}\cosh s_2\theta) \tag{1.101}$$

$$r_{12}(\theta) = \frac{1}{T_i}(\frac{1}{T}\cos s_1\theta + s_1 \sin s_1\theta - \frac{1}{T}\cosh s_2\theta + s_2 \sinh s_2\theta) \tag{1.102}$$

$$r_{22}(\theta) = s_2^2 \cos s_1\theta + s_1^2 \cosh s_2\theta \tag{1.103}$$

$$r_{32}(\theta) = \frac{k_0}{T T_i^2}(\frac{1}{s_1}\sin s_1\theta - \frac{1}{s_2}\sinh s_2\theta) \tag{1.104}$$

$$r_{42}(\theta) = \frac{k_0}{T T_i}(-\cos s_1\theta + \frac{1}{T s_1}\sin s_1\theta - \frac{1}{T s_2}\sinh s_2\theta + \cosh s_2\theta) \tag{1.105}$$

$$r_{13}(\theta) = \frac{k_0}{T T_i}(\cos s_1\theta - \cosh s_2\theta) \tag{1.106}$$

19

$$r_{23}(\theta) = \frac{k_0}{T}(-s_1 \sin s_1 \theta + \frac{1}{T} \cos s_1 \theta - s_2 \sinh s_2 \theta - \frac{1}{T} \cosh s_2 \theta) \quad (1.107)$$

$$r_{33}(\theta) = s_1^2 \cos s_1 \theta + \frac{s_1}{T} \sin s_1 \theta + s_2^2 \cosh s_2 \theta + \frac{s_2}{T} \sinh s_2 \theta \quad (1.108)$$

$$r_{43}(\theta) = \frac{k_0^2}{T^2 T_i}(\frac{1}{s_1} \sin s_1 \theta - \frac{1}{s_2} \sinh s_2 \theta) \quad (1.109)$$

$$r_{14}(\theta) = \frac{k_0}{T T_{i^2}}(-\frac{1}{s_1} \sin s_1 \theta + \frac{1}{s_2} \sinh s_2 \theta) \quad (1.110)$$

$$r_{24}(\theta) = \frac{k_0}{T T_i}(-\cos s_1 \theta - \frac{1}{T s_1} \sin s_1 \theta + \frac{1}{T s_2} \sinh s_2 \theta + \cosh s_2 \theta) \quad (1.111)$$

$$r_{34}(\theta) = \frac{1}{T_i}(\frac{1}{T} \cos s_1 \theta - s_1 \sin s_1 \theta - \frac{1}{T} \cosh s_2 \theta - s_2 \sinh s_2 \theta) \quad (1.112)$$

$$r_{44}(\theta) = s_2^2 \cos s_1 \theta + s_1^2 \cosh s_2 \theta \quad (1.113)$$

The solution of the differential equations (1.88) is given

$$\beta_{12}(\theta) = \frac{1}{s_1^2 + s_2^2}[r_{11}(\theta + r)\beta_{12}(-r) + r_{12}(\theta + r)\beta_{22}(-r) + r_{13}(\theta + r)\kappa_{12}(-r) - r_{14}(\theta + r)] \quad (1.114)$$

$$\beta_{22}(\theta) = \frac{1}{s_1^2 + s_2^2}[r_{21}(\theta + r)\beta_{12}(-r) + r_{22}(\theta + r)\beta_{22}(-r) + r_{23}(\theta + r)\kappa_{12}(-r) - r_{24}(\theta + r)] \quad (1.115)$$

$$\kappa_{12}(\theta) = \frac{1}{s_1^2 + s_2^2}[r_{31}(\theta + r)\beta_{12}(-r) + r_{32}(\theta + r)\beta_{22}(-r) + r_{33}(\theta + r)\kappa_{12}(-r) - r_{34}(\theta + r)] \quad (1.116)$$

$$\kappa_{22}(\theta) = \frac{1}{s_1^2 + s_2^2}[r_{41}(\theta + r)\beta_{12}(-r) + r_{42}(\theta + r)\beta_{22}(-r) + r_{43}(\theta + r)\kappa_{12}(-r) - r_{44}(\theta + r)] \quad (1.117)$$

The matrix α and the initial conditions $\beta_{12}(-r)$, $\beta_{22}(-r)$, $\kappa_{12}(-r)$ are obtained from the set of algebraic equations

$$\begin{cases} -\frac{2}{T}\alpha_{11} + \frac{2}{T_i}\alpha_{12} = -1 \\ -\frac{2}{T}\alpha_{12} + \frac{2}{T_i}\alpha_{22} + \kappa_{12}(-r) = 0 \\ -\frac{2k_0}{T}\alpha_{11} - \beta_{12}(-r) = 0 \\ -\frac{2k_0}{T}\alpha_{12} - \beta_{22}(-r) = 0 \\ q_{11}\beta_{12}(-r) + q_{12}\beta_{22}(-r) + q_{13}\kappa_{12}(-r) = q_{14} \\ q_{21}\beta_{12}(-r) + q_{22}\beta_{22}(-r) + q_{23}\kappa_{12}(-r) = q_{24} \end{cases} \tag{1.118}$$

where

$$q_{11} = (s_1^2 - \frac{k_0}{TT_i})\cos\frac{s_1 r}{2} - \frac{s_1}{T}\sin\frac{s_1 r}{2} + (s_2^2 + \frac{k_0}{TT_i})\cosh\frac{s_2 r}{2} - \frac{s_2}{T}\sinh\frac{s_2 r}{2} \tag{1.119}$$

$$q_{12} = \frac{1}{TT_i}\cos\frac{s_1 r}{2} + \frac{s_1^2 - \frac{k_0}{TT_i}}{T_i s_1}\sin\frac{s_1 r}{2} - \frac{1}{TT_i}\cosh\frac{s_2 r}{2} + \frac{s_2^2 + \frac{k_0}{TT_i}}{T_i s_2}\sinh\frac{s_2 r}{2} \tag{1.120}$$

$$q_{13} = (\frac{k_0}{TT_i} - s_1^2)\cos\frac{s_1 r}{2} - \frac{s_1}{T}\sin\frac{s_1 r}{2} - (\frac{k_0}{TT_i} + s_2^2)\cosh\frac{s_2 r}{2} - \frac{s_2}{T}\sinh\frac{s_2 r}{2} \tag{1.121}$$

$$q_{14} = -\frac{1}{TT_i}\cos\frac{s_1 r}{2} + \frac{s_1^2 - \frac{k_0}{TT_i}}{T_i s_1}\sin\frac{s_1 r}{2} + \frac{1}{TT_i}\cosh\frac{s_2 r}{2} + \frac{s_2^2 + \frac{k_0}{TT_i}}{T_i s_2}\sinh\frac{s_2 r}{2} \tag{1.122}$$

$$q_{21} = -\frac{k_0}{T^2}\cos\frac{s_1 r}{2} - \frac{k_0(s_1^2 + \frac{k_0}{TT_i})}{T s_1}\sin\frac{s_1 r}{2} + \frac{k_0}{T^2}\cosh\frac{s_2 r}{2} - \frac{k_0(s_2^2 - \frac{k_0}{TT_i})}{T s_2}\sinh\frac{s_2 r}{2} \tag{1.123}$$

$$q_{22} = (s_2^2 + \frac{k_0}{TT_i})\cos\frac{s_1 r}{2} - \frac{k_0}{T^2 T_i s_1}\sin\frac{s_1 r}{2} + (s_1^2 - \frac{k_0}{TT_i})\cosh\frac{s_2 r}{2} + \frac{k_0}{T^2 T_i s_2}\sinh\frac{s_2 r}{2} \tag{1.124}$$

$$q_{23} = \frac{k_0}{T^2}\cos\frac{s_1 r}{2} - \frac{k_0(s_1^2 + \frac{k_0}{TT_i})}{T s_1}\sin\frac{s_1 r}{2} - \frac{k_0}{T^2}\cosh\frac{s_2 r}{2} - \frac{k_0(s_2^2 - \frac{k_0}{TT_i})}{T s_2}\sinh\frac{s_2 r}{2} \tag{1.125}$$

$$q_{24} = -(s_2^2 + \frac{k_0}{TT_i})\cos\frac{s_1 r}{2} - \frac{k_0}{T^2 T_i s_1}\sin\frac{s_1 r}{2} + (-s_1^2 + \frac{k_0}{TT_i})\cosh\frac{s_2 r}{2} + \frac{k_0}{T^2 T_i s_2}\sinh\frac{s_2 r}{2} \tag{1.126}$$

The solution of the set of the equations (1.118) is given

$$\alpha_{11} = \frac{1}{M}[(k_0 + \frac{T}{T_i})(s_1^2 + s_2^2)\cos\frac{s_1 r}{2}\cosh\frac{s_2 r}{2} + \frac{k_0(T - k_0 T_i)(s_1^2 + s_2^2)}{T_i^2 T s_1 s_2}\sin\frac{s_1 r}{2}\sinh\frac{s_2 r}{2} +$$

$$+ \frac{(s_2^2 - \frac{k_0}{TT_i})[1 + 2k_0^2 + k_0 TT_i(s_1^2 + s_2^2)]}{T_i s_2}\cos\frac{s_1 r}{2}\sinh\frac{s_2 r}{2} +$$

$$+ \frac{(s_1^2 + \frac{k_0}{TT_i})[1 + 2k_0^2 - k_0 TT_i(s_1^2 + s_2^2)]}{T_i s_1}\sin\frac{s_1 r}{2}\cosh\frac{s_2 r}{2}] \tag{1.127}$$

$$\alpha_{12} = \frac{1}{M}[(s_1^2 + s_2^2)\cos\frac{s_1 r}{2}\cosh\frac{s_2 r}{2} + \frac{k_0(s_1^2 + s_2^2)}{TT_i s_1 s_2}\sin\frac{s_1 r}{2}\sinh\frac{s_2 r}{2} +$$

$$+ \frac{(s_2^2 - \frac{k_0}{TT_i})(1 + 2k_0^2)}{Ts_2}\cos\frac{s_1 r}{2}\sinh\frac{s_2 r}{2} + \frac{(s_1^2 + \frac{k_0}{TT_i})(1 + 2k_0^2)}{Ts_1}\sin\frac{s_1 r}{2}\cosh\frac{s_2 r}{2}] \qquad (1.128)$$

$$\alpha_{22} = \frac{1}{M}[\frac{(k_0 T + k_0^2 T_i + T_i)(s_1^2 + s_2^2)}{T}\cos\frac{s_1 r}{2}\cosh\frac{s_2 r}{2} +$$

$$+ \frac{T_i s_2^2(s_2^2 + \frac{k_0^2}{T^2}) + \frac{k_0}{T}(s_1^2 - \frac{k_0^2}{T^2})}{s_2}\cos\frac{s_1 r}{2}\sinh\frac{s_2 r}{2} +$$

$$+ \frac{T_i s_1^2(-s_1^2 + \frac{k_0^2}{T^2}) + \frac{k_0}{T}(s_2^2 + \frac{k_0^2}{T^2})}{s_1}\sin\frac{s_1 r}{2}\cosh\frac{s_2 r}{2} +$$

$$+ \frac{k_0(-k_0 T + k_0^2 T_i + T_i)(s_1^2 + s_2^2)}{T^2 T_i s_1 s_2}\sin\frac{s_1 r}{2}\sinh\frac{s_2 r}{2}] \qquad (1.129)$$

$$\beta_{12}(-r) = -\frac{2k_0}{TM}[(k_0 + \frac{T}{T_i})(s_1^2 + s_2^2)\cos\frac{s_1 r}{2}\cosh\frac{s_2 r}{2} +$$

$$+ \frac{(s_2^2 - \frac{k_0}{TT_i})[1 + 2k_0^2 + k_0 TT_i(s_1^2 + s_2^2)]}{T_i s_2}\cos\frac{s_1 r}{2}\sinh\frac{s_2 r}{2} +$$

$$+ \frac{(s_1^2 + \frac{k_0}{TT_i})[1 + 2k_0^2 - k_0 TT_i(s_1^2 + s_2^2)]}{T_i s_1}\sin\frac{s_1 r}{2}\cosh\frac{s_2 r}{2} +$$

$$+ \frac{k_0(T - k_0 T_i)(s_1^2 + s_2^2)}{T_i^2 T s_1 s_2}\sin\frac{s_1 r}{2}\sinh\frac{s_2 r}{2}] \qquad (1.130)$$

$$\beta_{22}(-r) = -\frac{2k_0}{TM}[(s_1^2 + s_2^2)\cos\frac{s_1 r}{2}\cosh\frac{s_2 r}{2} + \frac{k_0(s_1^2 + s_2^2)}{TT_i s_1 s_2}\sin\frac{s_1 r}{2}\sinh\frac{s_2 r}{2} +$$

$$+ \frac{(s_2^2 - \frac{k_0}{TT_i})(1 + 2k_0^2)}{Ts_2}\cos\frac{s_1 r}{2}\sinh\frac{s_2 r}{2} + \frac{(s_1^2 + \frac{k_0}{TT_i})(1 + 2k_0^2)}{Ts_1}\sin\frac{s_1 r}{2}\cosh\frac{s_2 r}{2}] \qquad (1.131)$$

$$\kappa_{12}(-r) = \frac{2}{TM}[-\frac{k_0(k_0 T_i + T)(s_1^2 + s_2^2)}{T_i}\cos\frac{s_1 r}{2}\cosh\frac{s_2 r}{2} +$$

$$+ [(k_0 T_i + T)s_1 s_2^2 - \frac{k_0 s_1}{TT_i}(T - k_0 T_i)]\sin\frac{s_1 r}{2}\cosh\frac{s_2 r}{2} +$$

$$+ [\frac{k_0 s_2(k_0 T_i - T)}{TT_i} - (k_0 T_i + T)s_1^2 s_2]\cos\frac{s_1 r}{2}\sinh\frac{s_2 r}{2} +$$

$$+T(T-k_0T_i)(s_1^2+s_2^2)s_1s_2\sin\frac{s_1r}{2}\sinh\frac{s_2r}{2}] \tag{1.132}$$

where

$$M=-\frac{2k_0}{T}(s_1^2+s_2^2)[-\cos\frac{s_1r}{2}\cosh\frac{s_2r}{2}+\frac{k_0}{TT_is_1s_2}\sin\frac{s_1r}{2}\sinh\frac{s_2r}{2}+$$

$$+(\frac{k_0}{T_is_2}-Ts_2)\cos\frac{s_1r}{2}\sinh\frac{s_2r}{2}+(\frac{k_0}{T_is_1}+Ts_1)\sin\frac{s_1r}{2}\cosh\frac{s_2r}{2}] \tag{1.133}$$

According to the formula (1.89) the value of the index is given

$$J=V(y(0),y(\theta)) \tag{1.134}$$

After calculations one obtains

$$J=x_0^2\alpha_{11}-2x_0z_0\alpha_{12}+z_0^2\alpha_{22}-\frac{x_0z_0}{s_1^2+s_2^2}[(s_1\sin s_1r+\frac{1}{T}\cos s_1r+s_2\sinh s_2r+$$

$$-\frac{1}{T}\cosh s_2r)\beta_{12}(-r)+\frac{1}{TT_i}(\frac{1}{s_1}\sin s_1r-\frac{1}{s_2}\sinh s_2r-T\cos s_1r+T\cosh s_2r)\beta_{22}(-r)+$$

$$+\frac{k_0}{TT_i}(\frac{1}{s_1}\sin s_1r-\frac{1}{s_2}\sinh s_2r)\kappa_{12}(-r)-\frac{k_0}{TT_i^2}(\frac{1}{s_1^2}\cos s_1r+\frac{1}{s_2^2}\cosh s_2r-\frac{1}{s_1^2}-\frac{1}{s_2^2})]+$$

$$+\frac{z_0^2k_0r}{T^2T_i(s_1^2+s_2^2)}[k_0(\frac{1}{s_2}\sinh s_2r-\frac{1}{s_1}\sin s_1r)\beta_{12}(-r)+\frac{k_0}{T_i}(\frac{1}{s_1^2}\cos s_1r+$$

$$+\frac{1}{s_2^2}\cosh s_2r)\beta_{22}(-r)+T_i(\cos s_1r-Ts_1\sin s_1r-\cosh s_2r-Ts_2\sinh s_2r)\kappa_{12}(-r)+$$

$$+\frac{1}{s_1}\sin s_1r+T\cos s_1r-\frac{1}{s_2}\sinh s_2r-T\cosh s_2r] \tag{1.135}$$

Chapter 2

A linear system with both lumped and distributed retarded type time delay.

2.1 A mathematical model of a linear system with both lumped and distributed retarded type time delay

Let us consider the linear system with both lumped and distributed delay, whose dynamics is described by equation

$$\begin{cases} \frac{dx(t)}{dt} = Ax(t) + Bx_t(-r) + \int_{-r}^{0} Gx_t(\theta)d\theta \\ x(t_0) = x_0 \in \mathbb{R}^n \\ x_{t_0} = \Phi \in L^2([-r,0),\mathbb{R}^n) \end{cases} \qquad (2.1)$$

where A, B, $G \in \mathbb{R}^{n \times n}$, $x(t) \in \mathbb{R}^n$, $x_t(\theta) = x(t+\theta)$, $\theta \in [-r,0)$, $x_t \in W^{1,2}([-r,0),\mathbb{R}^n)$ for $t \geq t_0$ and $r > 0$.

$L^2([-r,0),\mathbb{R}^n)$ is a space of a Lebesgue square integrable functions on interval $[-r,0)$ with values in $\mathbb{R}^n$.

The state of a system (2.1) is a vector

$$S(t) = \begin{bmatrix} x(t) \\ x_t \end{bmatrix} \qquad (2.2)$$

where $x(t) \in \mathbb{R}^n$, $x_t \in L^2([-r,0),\mathbb{R}^n)$ for $t \geq t_0$

The state space is defined by a formula

$$X = \mathbb{R}^n \times L^2([-r,0),\mathbb{R}^n) \qquad (2.3)$$

$S = 0$ is an equilibrium point of the system (2.1).

In a parametric optimization problem is used an integral quadratic performance index of quality

$$J = \int_{t_0}^{\infty} x^T(t)Wx(t)dt \qquad (2.4)$$

where $W \in \mathbb{R}^{n \times n}$ is a positive definite matrix.

2.2 Determination of the Lyapunov functional

Let us consider a quadratic functional on X, given by a formula

$$V(x(t),x_t) = x^T(t)\alpha x(t) + \int_{-r}^{0} x^T(t)\beta(\theta)x_t(\theta)d\theta +$$

$$+ \int_{-r}^{0} \int_{\theta}^{0} x_t^T(\theta)\delta(\theta,\sigma)x_t(\sigma)d\sigma d\theta \qquad (2.5)$$

for $t \geq t_0$, where $\alpha \in \mathbb{R}^{n \times n}$, $\beta \in C^1([-r,0],\mathbb{R}^{n \times n})$, $\delta \in C^1(\Omega,\mathbb{R}^{n \times n})$
$\Omega = \{(\theta,\sigma): \theta \in [-r,0], \sigma \in [\theta,0]\}$ C^1 is a space of continuous functions with continuous derivative.

Conjecture 14. *It is given a procedure of determination of the functional (2.5) coefficients to obtain the Lyapunov functional.*

The time derivative of the functional (2.5) on the trajectory of the system (2.1) is computed. It is taken the following procedure. One computes the time derivative of each term of the right-hand-side of the formula (2.5) and one substitutes in place of $\frac{dx(t)}{dt}$ and $\frac{\partial x_t(\theta)}{\partial t}$ the following terms

$$\frac{dx(t)}{dt} = Ax(t) + Bx_t(-r) + \int_{-r}^{0} Gx_t(\theta)d\theta \qquad (2.6)$$

$$\frac{\partial x_t(\theta)}{\partial t} = \frac{\partial x_t(\theta)}{\partial \theta} \qquad (2.7)$$

In such a manner one attains

$$\frac{dV(x(t),x_t)}{dt} = x^T(t)\left[A^T\alpha + \alpha A + \frac{\beta(0) + \beta^T(0)}{2}\right]x(t) +$$

$$+ x_t^T(-r)\left[2B^T\alpha - \beta^T(-r)\right]x(t) +$$

$$+ \int_{-r}^{0} x^T(t)\left[2\alpha G + A^T\beta(\theta) - \frac{d\beta(\theta)}{d\theta} + \delta^T(\theta,0) + \delta(0,\theta)\right]x_t(\theta)d\theta +$$

$$+ \int_{-r}^{0} x_t^T(-r)[B^T\beta(\theta) - \delta^T(\theta,-r) - \delta(-r,\theta)]x_t(\theta)d\theta +$$

$$- \int_{-r}^{0}\int_{-r}^{0} x_t^T(\theta)\left[\frac{\partial\delta(\theta,\sigma)}{\partial\theta} + \frac{\partial\delta(\theta,\sigma)}{\partial\sigma} - G^T\beta(\sigma)\right]x_t(\sigma)d\sigma d\theta \tag{2.8}$$

Assume that the time derivative of the Lyapunov functional V is given as a quadratic form

$$\frac{dV(x(t),x_t)}{dt} \equiv -x^T(t)Wx(t) \tag{2.9}$$

for $t \geq t_0$, where $W \in \mathbb{R}^{n\times n}$ is a positive definite matrix.

When is known the Lyapunov functional and the relationship (2.9) holds , one can easily determine the value of a square indicator of the quality of the parametric optimization, because

$$J = \int_{t_0}^{\infty} x^T(t)Wx(t)dt = V(x(t_0),\Phi) \tag{2.10}$$

From equations (2.8) and (2.9) one obtains the system of equations (2.11) to (2.15)

$$A^T\alpha + \alpha A + \frac{\beta(0) + \beta^T(0)}{2} = -W \tag{2.11}$$

$$2B^T\alpha - \beta^T(-r) = 0 \tag{2.12}$$

$$A^T\beta(\theta) - \frac{d\beta(\theta)}{d\theta} + \delta(0,\theta) + \delta^T(\theta,0) + 2\alpha G = 0 \tag{2.13}$$

$$B^T\beta(\theta) - \delta(-r,\theta) - \delta^T(\theta,-r) = 0 \tag{2.14}$$

$$\frac{\partial\delta(\theta,\sigma)}{\partial\theta} + \frac{\partial\delta(\theta,\sigma)}{\partial\sigma} - G^T\beta(\sigma) = 0 \tag{2.15}$$

for $\theta \in [-r,0]$, $\sigma \in [-r,0]$.

Let us consider that the solution of the equation (2.15) is as below

$$\delta(\theta,\sigma) = \varphi(\theta-\sigma) + \varphi^T(\sigma-\theta) + \int_{0}^{\sigma} G^T\beta(\xi)d\xi \tag{2.16}$$

where $\varphi \in C^1([-r,r],\mathbb{R}^{n\times n})$.

From equation (2.16) one obtains

$$\delta^T(\theta,0) = \varphi(-\theta) + \varphi^T(\theta) \tag{2.17}$$

26

$$\delta(0,\theta) = \varphi(-\theta) + \varphi^T(\theta) + \int_0^\theta G^T \beta(\xi) d\xi \tag{2.18}$$

$$\delta(-r,\theta) = \varphi(-\theta-r) + \varphi^T(\theta+r) + \int_0^\theta G^T \beta(\xi) d\xi \tag{2.19}$$

$$\delta^T(\theta,-r) = \varphi^T(\theta+r) + \varphi(-\theta-r) + \int_0^{-r} \beta^T(\xi) G d\xi \tag{2.20}$$

One puts (2.17) and (2.18) into (2.13), and one gets the formula

$$-\frac{d\beta(\theta)}{d\theta} + A^T \beta(\theta) + 2\varphi^T(\theta) + 2\varphi(-\theta) + \int_0^\theta G^T \beta(\xi) d\xi + 2\alpha G = 0 \tag{2.21}$$

for $\theta \in [-r,0]$.

One substitutes (2.19) and (2.20) into (2.14). After some calculations one obtains

$$2\varphi^T(\theta) + 2\varphi(-\theta) = \beta^T(-\theta-r)B - \int_0^{-\theta-r} \beta^T(\xi) G d\xi - \int_0^{-r} G^T \beta(\xi) d\xi \tag{2.22}$$

One puts (2.22) into (2.21). After some calculations one attains

$$\frac{d\beta(\theta)}{d\theta} = A^T \beta(\theta) + \beta^T(-\theta-r)B + \int_{-r}^\theta \beta^T(-\xi-r) G d\xi + \int_{-r}^\theta G^T \beta(\xi) d\xi + 2\alpha G \tag{2.23}$$

A new function is introduced

$$\kappa(\theta) = \beta(-\theta-r) \tag{2.24}$$

for $\theta \in [-r,0]$.

Now the formula (2.23) can be written in the form

$$\frac{d\beta(\theta)}{d\theta} = A^T \beta(\theta) + \kappa^T(\theta)B + \int_{-r}^\theta \kappa^T(\xi) G d\xi + \int_{-r}^\theta G^T \beta(\xi) d\xi + 2\alpha G \tag{2.25}$$

One calculates the derivative of the function κ given by the formula (2.24). The relation (2.25) was taken into account

$$\frac{d\kappa(\theta)}{d\theta} = -A^T \kappa(\theta) - \beta^T(\theta)B + \int_0^\theta G^T \kappa(\xi) d\xi + \int_0^\theta \beta^T(\xi) G d\xi - 2\alpha G \tag{2.26}$$

27

One introduces two new functions

$$\eta(\theta) = A^T \beta(\theta) + \kappa^T(\theta)B + \int_{-r}^{\theta} \kappa^T(\xi)Gd\xi + \int_{-r}^{\theta} G^T\beta(\xi)d\xi + 2\alpha G \tag{2.27}$$

$$\vartheta(\theta) = -A^T \kappa(\theta) - \beta^T(\theta)B + \int_{0}^{\theta} G^T\kappa(\xi)d\xi + \int_{0}^{\theta} \beta^T(\xi)Gd\xi - 2\alpha G \tag{2.28}$$

Functions η and ϑ are not independent. It is easy to check that they are linked by formula

$$\eta(-\theta - r) = -\vartheta(\theta) \tag{2.29}$$

for $\theta \in [-r, 0]$.

From equations (2.25) and (2.27) it results that

$$\frac{d\beta(\theta)}{d\theta} = \eta(\theta) \tag{2.30}$$

From equations (2.26) and (2.28) it results that

$$\frac{d\kappa(\theta)}{d\theta} = \vartheta(\theta) \tag{2.31}$$

The derivatives of (2.27) and (2.28) are computed. Upon taking the relations (2.30) and (2.31) into account, one gets the formulas

$$\frac{d\eta(\theta)}{d\theta} = A^T \eta(\theta) + \vartheta^T(\theta)B + G^T\beta(\theta) + \kappa^T(\theta)G \tag{2.32}$$

$$\frac{d\vartheta(\theta)}{d\theta} = -A^T \vartheta(\theta) - \eta^T(\theta)B + G^T\kappa(\theta) + \beta^T(\theta)G \tag{2.33}$$

One obtains the system of differential equations

$$\begin{cases} \frac{d\beta(\theta)}{d\theta} = \eta(\theta) \\ \frac{d\kappa(\theta)}{d\theta} = \vartheta(\theta) \\ \frac{d\eta(\theta)}{d\theta} = A^T \eta(\theta) + \vartheta^T(\theta)B + G^T\beta(\theta) + \kappa^T(\theta)G \\ \frac{d\vartheta(\theta)}{d\theta} = -A^T \vartheta(\theta) - \eta^T(\theta)B + G^T\kappa(\theta) + \beta^T(\theta)G \end{cases} \tag{2.34}$$

for $\theta \in [-r, 0]$.

The solution of the differential equations (2.34) satisfies the conditions

$$\beta(\theta)\,|_{\theta=-\frac{r}{2}} = \kappa(\theta)\,|_{\theta=-\frac{r}{2}} \tag{2.35}$$

$$\eta(\theta)\,|_{\theta=-\frac{r}{2}} = -\vartheta(\theta)\,|_{\theta=-\frac{r}{2}} \tag{2.36}$$

Formula (2.35) was obtained from (2.24) and formula (2.36) from (2.29).

Now will be obtained the initial conditions of the differential equations (2.34).

From equation (2.24) it results that

$$\kappa(-r) = \beta(0) \tag{2.37}$$

From equation (2.27) it results that

$$\eta(-r) = A^T \beta(-r) + \kappa^T(-r)B + 2\alpha G \tag{2.38}$$

Upon taking the relation (2.37) into account, equations (2.11) and (2.12) take the form

$$A^T \alpha + \alpha A + \frac{\kappa(-r) + \kappa^T(-r)}{2} = -W \tag{2.39}$$

$$2B^T \alpha - \beta^T(-r) = 0 \tag{2.40}$$

One obtains the system of algebraic equations

$$\begin{cases} A^T \alpha + \alpha A + \frac{\kappa(-r) + \kappa^T(-r)}{2} = -W \\ 2B^T \alpha - \beta^T(-r) = 0 \\ -\eta(-r) + A^T \beta(-r) + \kappa^T(-r)B + 2\alpha G = 0 \\ \beta(\theta)\,|_{\theta=-\frac{r}{2}} = \kappa(\theta)\,|_{\theta=-\frac{r}{2}} \\ \eta(\theta)\,|_{\theta=-\frac{r}{2}} = -\vartheta(\theta)\,|_{\theta=-\frac{r}{2}} \end{cases} \tag{2.41}$$

The set of algebraic equations (2.41) allows for determination of the matrix α and the initial conditions of the system of differential equations (2.34).

From equations (2.21) and (2.30) one obtains

$$\varphi^T(\theta) + \varphi(-\theta) = -\alpha G - \frac{1}{2}A^T \beta(\theta) + \frac{1}{2}\eta(\theta) - \frac{1}{2}\int_0^\theta G^T \beta(\xi)d\xi \tag{2.42}$$

Putting (2.42) into (2.16), one gets the matrix $\delta(\theta,\sigma)$

$$\delta(\theta,\sigma) = -\frac{1}{2}\beta^T(\theta-\sigma)A + \frac{1}{2}\eta^T(\theta-\sigma) - \frac{1}{2}\int_0^{\theta-\sigma}\beta^T(\xi)Gd\xi + \int_0^\sigma G^T \beta(\xi)d\xi - G^T \alpha \tag{2.43}$$

for $\theta \in [-r,0]$, $\sigma \in [-r,0]$.

In this way one obtained all coefficients of the functional (2.5). This coefficients depend on the matrices A , B and G of the system (2.1). The time derivative of the functional (2.5) is negative definite. When the matrices α, $\beta(\theta)$ and $\delta(\theta,\sigma)$ for $t \geq t_0, \theta \in [-r,0]$, $\sigma \in [\theta,0]$ are positive definite then the functional (2.5) becomes the Lyapunov functional.

2.3 An example

Let us consider the system described by equation

$$
\begin{cases}
\frac{dx(t)}{dt} = ax(t) + bx_t(-r) + \int_{-r}^0 gx_t(\theta)d\theta \\
x(t_0) = x_0 \in \mathbb{R} \\
x_{t_0} = \Phi \in W^{1,2}([-r,0),\mathbb{R})
\end{cases}
\tag{2.44}
$$

$t \geq t_0$, $x(t) \in \mathbb{R}$, $x_t(\theta) = x(t+\theta)$, $\theta \in [-r,0)$, $x_t \in W^{1,2}([-r,0),\mathbb{R})$, $a, b, g \in \mathbb{R}$, $r > 0$

The Lyapunov functional is defined by the formula

$$
V(x(t),x_t) = \alpha x^2(t) + \int_{-r}^0 x(t)\beta(\theta)x_t(\theta)d\theta + \int_{-r}^0 \int_{-r}^0 x_t(\theta)\delta(\theta,\sigma)x_t(\sigma)d\sigma d\theta
\tag{2.45}
$$

In a parametric optimization problem is used an integral quadratic performance index of quality

$$
J = \int_{t_0}^{\infty} wx^2(t)dt = V(x_0,\Phi)
\tag{2.46}
$$

The set of equations (2.34) becomes

$$
\begin{bmatrix}
\frac{d\beta(\theta)}{d\theta} \\
\frac{d\eta(\theta)}{d\theta} \\
\frac{d\kappa(\theta)}{d\theta} \\
\frac{d\vartheta(\theta)}{d\theta}
\end{bmatrix}
=
\begin{bmatrix}
0 & 1 & 0 & 0 \\
g & a & g & b \\
0 & 0 & 0 & 1 \\
g & -b & g & -a
\end{bmatrix}
\begin{bmatrix}
\beta(\theta) \\
\eta(\theta) \\
\kappa(\theta) \\
\vartheta(\theta)
\end{bmatrix}
\tag{2.47}
$$

The fundamental matrix of solutions of equation (1.88) is given by

$$
R(\theta) =
\begin{bmatrix}
r_{11}(\theta) & r_{12}(\theta) & r_{13}(\theta) & r_{14}(\theta) \\
r_{21}(\theta) & r_{22}(\theta) & r_{23}(\theta) & r_{24}(\theta) \\
r_{31}(\theta) & r_{32}(\theta) & r_{33}(\theta) & r_{34}(\theta) \\
r_{41}(\theta) & r_{42}(\theta) & r_{43}(\theta) & r_{44}(\theta)
\end{bmatrix}
\tag{2.48}
$$

where

$$
r_{11}(\theta) = 1 - \frac{g}{s^2} + \frac{1}{s^2(b^2 - ab - g)}[g(bs^2 + ag - bg)\theta +
$$

$$
- \frac{g}{2s}(g+bs)(s+a-b)\exp(s\theta) - \frac{g}{2s}(g-bs)(s-a+b)\exp(-s\theta)]
\tag{2.49}
$$

$$
r_{21}(\theta) = \frac{1}{s^2(b^2 - ab - g)}[g(bs^2 + ag - bg) - \frac{g}{2}(g+bs)(s+a-b)\exp(s\theta) +
$$

$$
+ \frac{g}{2}(g-bs)(s-a+b)\exp(-s\theta)]
\tag{2.50}
$$

$$r_{31}(\theta) = -\frac{g}{s^2} + \frac{1}{s^2(b^2 - ab - g)}[-g(bs^2 + ag - bg)\theta - \frac{g}{2s}(s^2 - as - g)(s + a - b)\exp(s\theta)+$$

$$-\frac{g}{2s}(s^2 + as - g)(s - a + b)\exp(-s\theta)] \tag{2.51}$$

$$r_{41}(\theta) = \frac{1}{s^2(b^2 - ab - g)}[-g(bs^2 + ag - bg) - \frac{g}{2}(s^2 - as - g)(s + a - b)\exp(s\theta)+$$

$$+\frac{g}{2}(s^2 + as - g)(s - a + b)\exp(-s\theta)] \tag{2.52}$$

$$r_{12}(\theta) = -\frac{a}{s^2} + \frac{g}{s^2}\theta + \frac{1}{s^2(b^2 - ab - g)}[-\frac{1}{2s}(g + bs)(a^2 + as - bs - ab + g)\exp(s\theta)+$$

$$+\frac{1}{2s}(g - bs)(a^2 - as + bs - ab + g)\exp(-s\theta)] \tag{2.53}$$

$$r_{22}(\theta) = \frac{g}{s^2} + \frac{1}{s^2(b^2 - ab - g)}[-\frac{1}{2}(g + bs)(a^2 + as - bs - ab + g)\exp(s\theta)+$$

$$-\frac{1}{2}(g - bs)(a^2 - as + bs - ab + g)\exp(-s\theta)] \tag{2.54}$$

$$r_{32}(\theta) = \frac{h}{s^2} - \frac{g}{s^2}\theta + \frac{1}{s^2(b^2 - ab - g)}[-\frac{1}{2s}(s^2 - as - g)(a^2 + as - bs - ab + g)\exp(s\theta)+$$

$$+\frac{1}{2s}(s^2 + as - g)(a^2 - as + bs - ab + g)\exp(-s\theta)] \tag{2.55}$$

$$r_{42}(\theta) = -\frac{g}{s^2} + \frac{1}{s^2(b^2 - ab - g)}[-\frac{1}{2}(s^2 - as - g)(a^2 + as - bs - ab + g)\exp(s\theta)+$$

$$-\frac{1}{2}(s^2 + as - g)(a^2 - as + bs - ab + g)\exp(-s\theta)] \tag{2.56}$$

$$r_{13}(\theta) = -\frac{g}{s^2} + \frac{1}{s^2(b^2 - ab - g)}[g(bs^2 + ag - bg)\theta - \frac{g}{2s}(g + bs)(s + a - b)\exp(s\theta)+$$

$$-\frac{g}{2s}(g - bs)(s - a + b)\exp(-s\theta)] \tag{2.57}$$

$$r_{23}(\theta) = \frac{1}{s^2(b^2 - ab - g)}[g(bs^2 + ag - bg) - \frac{g}{2}(g + bs)(s + a - b)\exp(s\theta)+$$

$$+\frac{g}{2}(g - bs)(s - a + b)\exp(-s\theta)] \tag{2.58}$$

$$r_{33}(\theta) = 1 - \frac{g}{s^2} + \frac{1}{s^2(b^2 - ab - g)}[-g(bs^2 + ag - bg)\theta +$$

$$-\frac{g}{2s}(s^2 - as - g)(s + a - b)\exp(s\theta) - \frac{g}{2s}(s^2 + as - g)(s - a + b)\exp(-s\theta)] \tag{2.59}$$

$$r_{43}(\theta) = \frac{1}{s^2(b^2 - ab - g)}[-g(bs^2 + ag - bg) - \frac{g}{2}(s^2 - as - g)(s + a - b)\exp(s\theta) +$$

$$+\frac{g}{2}(s^2 + as - g)(s - a + b)\exp(-s\theta)] \tag{2.60}$$

$$r_{14}(\theta) = -\frac{b}{s^2} - \frac{g}{s^2}\theta + \frac{1}{2s^3}(g + bs)\exp(s\theta) - \frac{1}{2s^3}(g - bs)\exp(-s\theta) \tag{2.61}$$

$$r_{24}(\theta) = -\frac{g}{s^2} + \frac{1}{2s^2}(g + bs)\exp(s\theta) + \frac{1}{2s^2}(g - bs)\exp(-s\theta) \tag{2.62}$$

$$r_{34}(\theta) = \frac{a}{s^2} + \frac{g}{s^2}\theta + \frac{1}{2s^3}(s^2 - as - g)\exp(s\theta) - \frac{1}{2s^3}(s^2 + as - g)\exp(-s\theta) \tag{2.63}$$

$$r_{44}(\theta) = \frac{g}{s^2} + \frac{1}{2s^2}(s^2 - as - g)\exp(s\theta) + \frac{1}{2s^2}(s^2 + as - g)\exp(-s\theta) \tag{2.64}$$

where

$$s = \sqrt{a^2 - b^2 + 2g} \tag{2.65}$$

The solution of the set of equations (2.47) is given

$$\beta(\theta) = r_{11}(\theta + r)\beta(-r) + r_{12}(\theta + r)\eta(-r) + r_{13}(\theta + r)\kappa(-r) + r_{14}(\theta + r)\vartheta(-r) \tag{2.66}$$

$$\eta(\theta) = r_{21}(\theta + r)\beta(-r) + r_{22}(\theta + r)\eta(-r) + r_{23}(\theta + r)\kappa(-r) + r_{24}(\theta + r)\vartheta(-r) \tag{2.67}$$

$$\kappa(\theta) = r_{31}(\theta + r)\beta(-r) + r_{32}(\theta + r)\eta(-r) + r_{33}(\theta + r)\kappa(-r) + r_{34}(\theta + r)\vartheta(-r) \tag{2.68}$$

$$\vartheta(\theta) = r_{41}(\theta + r)\beta(-r) + r_{42}(\theta + r)\eta(-r) + r_{43}(\theta + r)\kappa(-r) + r_{44}(\theta + r)\vartheta(-r) \tag{2.69}$$

Now will be given the formulas for determination of the set of initial conditions of equation (2.47) and the coefficient α

$$
\begin{cases}
2a\alpha + \kappa(-r) = -w \\
2b\alpha - \beta(-r) = 0 \\
-\eta(-r) + a\beta(-r) + b\kappa(-r) + 2g\alpha = 0 \\
\beta(\theta)\,|_{\theta=-\frac{r}{2}} = \kappa(\theta)\,|_{\theta=-\frac{r}{2}} \\
\eta(\theta)\,|_{\theta=-\frac{r}{2}} = -\vartheta(\theta)\,|_{\theta=-\frac{r}{2}}
\end{cases}
\tag{2.70}
$$

The set of algebraic equations (2.70) can be written in the equivalent form

$$
\kappa(-r) = -w - 2a\alpha
\tag{2.71}
$$

$$
\beta(-r) = 2b\alpha
\tag{2.72}
$$

$$
\eta(-r) = (2g - bw)\alpha
\tag{2.73}
$$

$$
2p_{11}\alpha + p_{12}\vartheta(-r) = p_{13}w
\tag{2.74}
$$

$$
2p_{21}\alpha + p_{22}\vartheta(-r) = p_{23}w
\tag{2.75}
$$

where

$$
p_{11} = (s^2 - g)(a + b - gr) - \frac{g}{2s}(a^2 - b^2 - as - bs)\exp(-\frac{sr}{2}) +
$$
$$
+ \frac{g}{2s}(a^2 - b^2 + as + bs)\exp(\frac{sr}{2})
\tag{2.76}
$$

$$
p_{12} = -a - b + gr - \frac{1}{2s}(a^2 - b^2 - as - bs)\exp(-\frac{sr}{2}) +
$$
$$
+ \frac{1}{2s}(a^2 - b^2 + as + bs)\exp(\frac{sr}{2})
\tag{2.77}
$$

$$
p_{13} = -s^2 - ab - b^2 + agr + \frac{1}{2s}(bs^2 + b^2 s + abs + 2ag)\exp(-\frac{sr}{2}) +
$$
$$
- \frac{1}{2s}(bs^2 - b^2 s - abs + 2ag)\exp(\frac{sr}{2})
\tag{2.78}
$$

$$
p_{21} = \frac{gs}{2}(s - a + b)\exp(-\frac{sr}{2}) + \frac{gs}{2}(s + a - b)\exp(\frac{sr}{2})
\tag{2.79}
$$

$$
p_{22} = \frac{s}{2}(s - a + b)\exp(-\frac{sr}{2}) + \frac{s}{2}(s + a - b)\exp(\frac{sr}{2})
\tag{2.80}
$$

33

$$p_{23} = \frac{s}{2}(s^2 - a^2 + ab - bs)\exp(-\frac{sr}{2}) - \frac{s}{2}(s^2 - a^2 + ab - bs)\exp(\frac{sr}{2}) \qquad (2.81)$$

The formula for α is given

$$\alpha = \frac{1}{M}[-aw(a+b)(b^2 - ab - g) +$$

$$-\frac{w}{2}(s^2 - a^2 + ab - bs)(a^2 - b^2) + \frac{bsw}{2}(a+b)(s - a + b)\exp(-sr) +$$

$$+\frac{sw}{2}(-s^3 - a^3 - b^3 - 2b^2s - 2abs + a^2b + ab^2 - grs(s - a - b))\exp(-\frac{sr}{2}) +$$

$$+\frac{sw}{2}(-s^3 + a^3 + b^3 - 2as^2 - a^2b - ab^2 + grs(s + a - b))\exp(\frac{sr}{2})] \qquad (2.82)$$

where

$$M = s^3(a + b - gr)(s - a + b)\exp(-\frac{sr}{2}) + s^3(a + b - gr)(s + a - b)\exp(\frac{sr}{2}) \qquad (2.83)$$

Having the solution of equations (2.47) and the coefficient α one obtains $\delta(\theta, \sigma)$

$$\delta(\theta, \sigma) = -ga - \frac{1}{2}a\beta(\theta - \sigma) + \frac{1}{2}\eta(\theta - \sigma) - \frac{1}{2}\int\limits_0^{\theta - \sigma} g\beta(\xi)d\xi + \int\limits_0^{\sigma} g\beta(\xi)d\xi \qquad (2.84)$$

34

Chapter 3

A linear neutral system with k-non-commensurate delays

3.1 A mathematical model of a linear neutral system with k-non-commensurate delays

Let us consider a linear system with k-non-commensurate neutral type time delays, whose dynamics is described by the functional-differential equation (FDE)

$$\begin{cases} \frac{dx(t)}{dt} - \sum_{i=1}^{k} B_i \frac{dx(t-\tau_i)}{dt} = Ax(t) + \sum_{i=1}^{k} A_i x(t - \tau_i) \\ x(t_0) = x_0 \\ x(t_0 + \theta) = \Phi(\theta) \end{cases} \tag{3.1}$$

for $t \geq t_0, x(t) \in \mathbb{R}^n, A, A_i, B_i \in \mathbb{R}^{n \times n}, i = 1, ..., k, 0 \leq \tau_1 \leq ... \leq \tau_i \leq ... \leq \tau_k = r, \Phi \in W^{1,2}([-r,0), \mathbb{R}^n)$ for $\theta \in [-r, 0)$,

where $W^{1,2}([-r,0), \mathbb{R}^n)$ is a space of all absolutely continuous functions with derivatives in a space of Lebesgue square integrable functions on interval $[-r, 0)$ with values in $\mathbb{R}^n$.

Definition 15. The norm in $W^{1,2}([-r,0), \mathbb{R}^n)$ is defined by

$$\| \Phi \|_{W^{1,2}}^2 = \int\limits_{-r}^{0} \left(\| \Phi(t) \|_{\mathbb{R}^n}^2 + \| \frac{d\Phi(t)}{dt} \|_{\mathbb{R}^n}^2 \right) dt \tag{3.2}$$

where $\| \cdot \|_{\mathbb{R}^n}$ is an arbitrary norm in $\mathbb{R}^n$.

Definition 16. The function $x_t \in W^{1,2}([-r,0), \mathbb{R}^n)$ is called a shifted restriction of x to an interval $[t - r, t)$ and is defined by a formula

$$x_t(\theta) := x(t + \theta) \tag{3.3}$$

for $t \geq t_0, \theta \in [-r, 0)$.

Lemma 17. *There holds the relationship*

$$\frac{\partial x_t(\theta)}{\partial t} = \frac{\partial x_t(\theta)}{\partial \theta} \qquad (3.4)$$

Proof.

$$x_t(\theta) = x(t + \theta) \quad for\ t \geq t_0,\ \theta \in [-r, 0)$$

$$\frac{\partial x_t(\theta)}{\partial t} = \frac{\partial x(t + \theta)}{\partial t} = \frac{\partial x(\xi)}{\partial \xi}\frac{\partial \xi}{\partial t} = \frac{\partial x(\xi)}{\partial \xi} \quad for\ \xi = t + \theta$$

$$\frac{\partial x_t(\theta)}{\partial \theta} = \frac{\partial x(t + \theta)}{\partial \theta} = \frac{\partial x(\xi)}{\partial \xi}\frac{\partial \xi}{\partial \theta} = \frac{\partial x(\xi)}{\partial \xi} \quad for\ \xi = t + \theta$$

hence

$$\frac{\partial x_t(\theta)}{\partial t} = \frac{\partial x_t(\theta)}{\partial \theta}$$

$\square$

Using the formula (3.16) one can write the equation (3.1) in the form

$$\begin{cases} \frac{dx(t)}{dt} - \sum_{i=1}^{k} B_i \frac{dx_t(-\tau_i)}{dt} = Ax(t) + \sum_{i=1}^{k} A_i x_t(-\tau_i) \\ x(t_0) = x_0 \\ x_{t_0} = \Phi \in W^{1,2}([-r, 0), \mathbb{R}^n) \end{cases} \qquad (3.5)$$

for $t \geq t_0$.

The norm of an initial value (x_0, Φ) is given by

$$\| (x_0, \Phi) \| = \sqrt{\| x_0 \|_{\mathbb{R}^n}^2 + \| \Phi \|_{W^{1,2}}^2} \qquad (3.6)$$

The theorems of existence, continuous dependence and uniqueness of solutions of the equation (3.5) are given in [9].

A solution of the functional-differential equation (3.5) with initial value (x_0, Φ) or simply a solution through (x_0, Φ) is an absolutely continuous function defined for $t \geq t_0 - r$ with values in $\mathbb{R}^n$.

$$x(\cdot; (x_0, \Phi)) \in W^{1,2}([t_0 - r, \infty), \mathbb{R}^n) \qquad (3.7)$$

where $W^{1,2}([t_0 - r, \infty), \mathbb{R}^n)$ is a space of all absolutely continuous functions with derivatives in a space of Lebesgue square integrable functions on an interval $[t_0 - r, \infty)$ taking values in $\mathbb{R}^n$.

The function $x_t(\cdot; (x_0, \Phi)) \in W^{1,2}([-r, 0), \mathbb{R}^n)$ is a shifted restriction of $x(\cdot; (x_0, \Phi))$ to an interval $[t - r, t)$ and is given by a formula

$$x_t(\theta; (x_0, \Phi)) := x(t + \theta; (x_0, \Phi)) \qquad (3.8)$$

for $t \geq t_0,\ \theta \in [-r, 0)$

$$x_{t_0}(\cdot; (x_0, \Phi)) = \Phi \qquad (3.9)$$

Definition 18. The zero solution of (3.5) is **stable** if for any $\varepsilon > 0$ there is a $\delta > 0$ such that

$$\| (x_0, \Phi) \| < \delta$$

implies $\| x_t(\cdot; (x_0, \Phi)) \| < \varepsilon$ for $t \geq t_0$.

Definition 19. The zero solution of (3.5) is **asymptotically stable** if

$$\| x_t(\cdot; (x_0, \Phi)) \| \to 0$$

as $t \to \infty$.

Definition 20. The zero solution of (3.5) is **exponentially stable** if there exists an $\eta > 0$ and an positive constant M such that

$$\| x_t(\cdot; (x_0, \Phi)) \| \leq M e^{-\eta t} \| (x_0, \Phi) \| \tag{3.10}$$

for $t \geq t_0$.

Definition 21. The **difference equation** associated with (3.5) is given by

$$x(t) = \sum_{i=1}^{k} B_i x_t(-\tau_i) \tag{3.11}$$

for $t \geq t_0$.

Definition 22. The **spectrum** $\sigma(G)$ is a set of complex numbers λ for which a matrix $\lambda I - G$ is not invertible.

$$\sigma(G) = \{ \lambda \in \mathbb{C} : \det(\lambda I - G) = 0 \} \tag{3.12}$$

The arbitrary eigenvalue of the matrix G will be denoted as $\lambda(G)$.

Definition 23. The **spectral radius** of a matrix G is given by a form

$$\gamma(G) = \sup\{ | \lambda | : \lambda \in \sigma(G) \} \tag{3.13}$$

The eigenvalues of the neutral equation (3.5) for large modulus are asymptotically equal to the eigenvalues of the difference equation (3.11). The asymptotic stability of the difference equation (3.11) is the necessary condition of the asymptotic stability of the neutral equation (3.5).

According to the Theorem 9.6.1 in [10] a difference equation (3.11) for fixed rationally independent

$$0 < \tau_1 \leq \dots \leq \tau_j \leq \dots \leq \tau_k$$

is stable if

$$\sup\left\{ \gamma\left(\sum_{j=1}^{k} e^{i\theta_j} B_j \right) : \theta_j \in [0, 2\pi], 1 \leq j \leq k \right\} < 1 \tag{3.14}$$

where $\gamma\left(\sum_{j=1}^{k} e^{i\theta_j} B_j\right)$ is the spectral radius of a matrix $\sum_{j=1}^{k} e^{i\theta_j} B_j$.

If each B_j is a scalar then a difference equation is stable if and only if

$$\sum_{j=1}^{k} |B_j| < 1 \tag{3.15}$$

A new variable y, is defined by the formula

$$y(t) = x(t) - \sum_{i=1}^{k} B_i x_t(-\tau_i) \tag{3.16}$$

for $t \geq t_0$

Thus the equations (3.5) take the form

$$\begin{cases} \frac{dy(t)}{dt} = Ay(t) + \sum_{i=1}^{k}(A_i + AB_i)x_t(-\tau_i) \\ y(t) = x(t) - \sum_{i=1}^{k} B_i x_t(-\tau_i) \\ y(t_0) = x_0 - \sum_{i=1}^{k} B_i \Phi(-\tau_i) \\ x_{t_0} = \Phi \end{cases} \tag{3.17}$$

Let us assume that the matrices B_i for $i = 1, ..., k$ fulfill the condition (3.14).

State of the system (3.17) is a vector

$$S(t) = \begin{bmatrix} y(t) \\ x_t \end{bmatrix} \tag{3.18}$$

for $t \geq t_0$

The state space is defined by the formula

$$X = \mathbb{R}^n \times W^{1,2}([-r, 0), \mathbb{R}^n) \tag{3.19}$$

The norm in the state space X is defined by

$$\| S(t) \|_X^2 = \| y(t) \|_{\mathbb{R}^n}^2 + \| x_t \|_{W^{1,2}}^2 \tag{3.20}$$

for $t \geq t_0$.

3.2 A Lyapunov functional

Definition 24. A functional $V : X \to \mathbb{R}$ is **positive definite** if and only if it is continuous and $V(x) > 0$ for $x \neq 0$ and $V(0) = 0$.

A functional $V : X \to \mathbb{R}$ is **negative definite** if and only if it is continuous and $V(x) < 0$ for $x \neq 0$ and $V(0) = 0$.

Definition 25. The time derivative of the functional $V(y(t), x_t)$ at $(y(t_0), \Phi)$ on a trajectory of a system (3.17) is defined by a formula

$$\frac{dV(y(t_0), \Phi)}{dt} = \limsup_{h \to 0} \frac{1}{h} \left[V\left(y(t_0 + h), x_{t_0 + h}\right) - V\left(y(t_0), \Phi\right) \right] \quad (3.21)$$

Definition 26. The functional $V : X \to \mathbb{R}$ is called the **Lyapunov functional** if

1. V is positive definite

2. V is differentiable

3. A time derivative of V computed according to a formula (3.21) on the trajectory of the system (3.17) is negative definite

Existence of the Lyapunov functional for the system (3.17) is a sufficient condition for asymptotic stability of its zero solution.

When the system (3.17) is asymptotically stable

$$\int\limits_{t_0}^{\infty} \frac{dV(y(t), x_t)}{dt} dt =$$

$$= \lim_{t \to \infty} V(y(t), x_t) - \lim_{t \to t_0} V(y(t), x_t) =$$

$$= V\left(\lim_{t \to \infty} (y(t), x_t)\right) - V\left(\lim_{t \to t_0} (y(t), x_t)\right) =$$

$$= V(0) - V(y(t_0), \Phi) = -V(y(t_0), \Phi) \quad (3.22)$$

Assume that the time derivative of the Lyapunov functional V is given as a quadratic form

$$\frac{dV(y(t), x_t)}{dt} \equiv -y^T(t) W y(t) \quad (3.23)$$

for $t \geq t_0$, where $W \in \mathbb{R}^{n \times n}$ is a positive definite matrix.

It follows from (3.22) and (3.23) that

$$J = \int\limits_{t_0}^{\infty} y^T(t) W y(t) dt = V(y_0, \Phi) \quad (3.24)$$

Corollary 27. *If one constructs a Lyapunov functional such that its time derivative computed on the trajectory of the system (3.17) is given as a quadratic form (3.23) one can not only investigate the system (3.17) stability but also calculate a value of a square indicator of quality (3.24) of the parametric optimization problem.*

To calculate the value of the performance index (3.24), which is equal to the value of the Lyapunov functional at the initial state of the system (3.17), one needs a mathematical formula of that functional.

3.3 Determination of the Lyapunov functional

Let us consider a quadratic functional on X, given by a formula

$$V(y(t), x_t) = y^T(t)\alpha y(t) + \int_{-r}^{0} y^T(t)\beta(\theta)x_t(\theta)d\theta +$$

$$+ \int_{-r}^{0}\int_{\theta}^{0} x_t^T(\theta)\delta(\theta, \sigma)x_t(\sigma)d\sigma d\theta \qquad (3.25)$$

for $t \geq t_0$, where $\alpha \in \mathbb{R}^{n \times n}$, $\beta \in C^1([-r, 0], \mathbb{R}^{n \times n})$, $\delta \in C^1(\Omega, \mathbb{R}^{n \times n})$
$\Omega = \{(\theta, \sigma): \theta \in [-r, 0], \sigma \in [\theta, 0]\} C^1$ is a space of continuous functions with continuous derivative.

Conjecture 28. *It is given a procedure of determination of the functional (3.25) coefficients to obtain the Lyapunov functional.*

The time derivative of the functional (3.25) on the trajectory of the system (3.17) is computed. This time derivative is defined by the formula (3.21). It is taken the following procedure. One computes the time derivative of each term of the right-hand-side of the formula (3.25) and one substitutes in place of $\frac{dy(t)}{dt}$ and $\frac{\partial x_t(\theta)}{\partial t}$ the following terms

$$\frac{dy(t)}{dt} = Ay(t) + \sum_{i=1}^{k}(A_i + AB_i)x_t(-\tau_i) \qquad (3.26)$$

$$\frac{\partial x_t(\theta)}{\partial t} = \frac{\partial x_t(\theta)}{\partial \theta} \qquad (3.27)$$

In such a manner one attains

$$\frac{dV(y(t), x_t)}{dt} = y^T(t)\left[A^T\alpha + \alpha A + \beta(0)\right]y(t) +$$

$$+ y^T(t)\left[(\alpha + \alpha^T)(A_k + AB_k) + \beta(0)B_k - \beta(-r)\right]x_t(-r) +$$

$$+ \sum_{i=1}^{k-1} y^T(t)\left[(\alpha + \alpha^T)(A_i + AB_i) + \beta(0)B_i\right]x_t(-\tau_i) +$$

$$+ \int_{-r}^{0} y^T(t)\left[A^T\beta(\theta) - \frac{d\beta(\theta)}{d\theta} + \delta^T(\theta, 0)\right]x_t(\theta)d\theta +$$

$$+ \int_{-r}^{0} x_t^T(-r)[(A_k + AB_k)^T\beta(\theta) + B_k^T\delta^T(\theta, 0) - \delta(-r, \theta)]x_t(\theta)d\theta +$$

$$+ \int_{-r}^{0} \sum_{i=1}^{k-1} x_t^T(-\tau_i)[(A_i+AB_i)^T \beta(\theta)+B_i^T \delta^T(\theta,0)]x_t(\theta)d\theta +$$

$$- \int_{-r}^{0} \int_{\theta}^{0} x_t^T(\theta) \left[\frac{\partial \delta(\theta,\sigma)}{\partial \theta} + \frac{\partial \delta(\theta,\sigma)}{\partial \sigma} \right] x_t(\sigma)d\sigma d\theta \qquad (3.28)$$

for $t \geq t_0$.

The time derivative of the Lyapunov functional is negative definite, therefore one identifies the coefficients of the functional (3.25) such that the time derivative (3.28) satisfies the relationship (3.23).

From relations (3.23) and (3.28) one attains the set of equations

$$A^T \alpha + \alpha A + \beta(0) = -W \qquad (3.29)$$

$$(\alpha + \alpha^T)(A_i+AB_i)+\beta(0)B_i = 0 \ for \ i = 1,...,k-1 \qquad (3.30)$$

$$(\alpha + \alpha^T)(A_k+AB_k)+\beta(0)B_k - \beta(-r) = 0 \qquad (3.31)$$

$$A^T \beta(\theta) - \frac{d\beta(\theta)}{d\theta} + \delta^T(\theta,0) = 0 \qquad (3.32)$$

$$(A_k+AB_k)^T \beta(\theta)+B_k^T \delta^T(\theta,0) - \delta(-r,\theta) = 0 \qquad (3.33)$$

$$(A_i+AB_i)^T \beta(\theta)+B_i^T \delta^T(\theta,0) = 0 \quad for \ i = 1,...,k-1 \qquad (3.34)$$

$$\frac{\partial \delta(\theta,\sigma)}{\partial \theta} + \frac{\partial \delta(\theta,\sigma)}{\partial \sigma} = 0 \qquad (3.35)$$

for $\theta \in [-r,0]$, $\sigma \in [\theta,0]$.

3.3.1 The solution of equations (3.29) to (3.35) for $k = 1$.

This equations take a form

$$A^T \alpha + \alpha A + \beta(0) = -W \qquad (3.36)$$

$$(\alpha + \alpha^T)(A_1+AB_1)+\beta(0)B_1 - \beta(-r) = 0 \qquad (3.37)$$

$$A^T \beta(\theta) - \frac{d\beta(\theta)}{d\theta} + \delta^T(\theta,0) = 0 \qquad (3.38)$$

$$(A_1 + AB_1)^T \beta(\theta) + B_1^T \delta^T(\theta,0) - \delta(-r,\theta) = 0 \tag{3.39}$$

$$\frac{\partial \delta(\theta,\sigma)}{\partial \theta} + \frac{\partial \delta(\theta,\sigma)}{\partial \sigma} = 0 \tag{3.40}$$

for $\theta \in [-r,0]$, $\sigma \in [\theta,0]$.

The solution of equation (3.40) is as below

$$\delta(\theta,\sigma) = \varphi(\theta - \sigma) \tag{3.41}$$

for $\theta \in [-r,0]$, $\sigma \in [\theta,0]$, where $\varphi \in C^1([-r,r],\mathbb{R}^{n \times n})$, C^1 is a space of continuous functions with continuous derivative.

From equation (3.38) one determines the term

$$\delta^T(\theta,0) = \frac{d\beta(\theta)}{d\theta} - A^T \beta(\theta) = \varphi^T(\theta) \tag{3.42}$$

and one puts it into relation (3.39). After some calculations one gets

$$B_1^T \frac{d\beta(\theta)}{d\theta} + A_1^T \beta(\theta) - \delta(-r,\theta) = 0 \tag{3.43}$$

It follows from equation (3.42) that

$$\delta(-r,\theta) = \varphi(-r-\theta) = -\frac{d\beta^T(-r-\theta)}{d\theta} - \beta^T(-r-\theta)A \tag{3.44}$$

One puts the term (3.44) into (3.43) and one obtains

$$B_1^T \frac{d\beta(\theta)}{d\theta} + \frac{d\beta^T(-r-\theta)}{d\theta} = -A_1^T \beta(\theta) - \beta^T(-r-\theta)A \tag{3.45}$$

After putting in the relation (3.45) a new variable $-r-\theta$, instead of an independent variable θ, one attains an equation

$$B_1^T \frac{d\beta(-r-\theta)}{d\theta} + \frac{d\beta^T(\theta)}{d\theta} = A_1^T \beta(-r-\theta) + \beta^T(\theta)A \tag{3.46}$$

The set of differential equations are obtained

$$\begin{cases} B_1^T \frac{d\beta(\theta)}{d\theta} + \frac{d\beta^T(-r-\theta)}{d\theta} = -A_1^T \beta(\theta) - \beta^T(-r-\theta)A \\ B_1^T \frac{d\beta(-r-\theta)}{d\theta} + \frac{d\beta^T(\theta)}{d\theta} = A_1^T \beta(-r-\theta) + \beta^T(\theta)A \end{cases} \tag{3.47}$$

The new function is given

$$\kappa(\theta) = \beta(-r-\theta) \tag{3.48}$$

The set of equations (3.47) takes a form

$$\begin{cases} B_1^T \dfrac{d\beta(\theta)}{d\theta} + \dfrac{d\kappa^T(\theta)}{d\theta} = -A_1^T \beta(\theta) - \kappa^T(\theta)A \\ B_1^T \dfrac{d\kappa(\theta)}{d\theta} + \dfrac{d\beta^T(\theta)}{d\theta} = A_1^T \kappa(\theta) + \beta^T(\theta)A \end{cases} \qquad (3.49)$$

The set of equations (3.49) can be written in the form

$$\begin{cases} \dfrac{d\beta(\theta)}{d\theta} - B_1^T \dfrac{d\beta^T(\theta)}{d\theta} B_1 = A_1^T \beta(\theta) B_1 + A^T \beta(\theta) + \kappa^T(\theta)(AB_1 + A_1) \\ \dfrac{d\kappa(\theta)}{d\theta} - B_1^T \dfrac{d\kappa^T(\theta)}{d\theta} B_1 = -\beta^T(\theta)(A_1 + AB_1) - A^T \kappa(\theta) - A_1^T \kappa(\theta) B_1 \end{cases} \qquad (3.50)$$

To obtain a solution of the equations (3.50) one needs the initial values $\beta(-r)$ and $\kappa(-r)$. The equation (3.48) implies that

$$\kappa(-r) = \beta(0) \qquad (3.51)$$

$$\beta(\theta)\,|_{\theta=-\frac{r}{2}} = \kappa(\theta)\,|_{\theta=-\frac{r}{2}} \qquad (3.52)$$

The equations (3.36) and (3.37) take a form

$$A^T \alpha + \alpha A + \kappa(-r) = -W \qquad (3.53)$$

$$\left(\alpha + \alpha^T\right)(A_1 + AB_1) + \kappa(-r)B_1 - \beta(-r) = 0 \qquad (3.54)$$

The set of algebraic equations (3.52) to (3.54) enables to obtain the matrix α and the initial conditions of the ordinary differential equations (3.50).

3.3.2 The solution of equations (3.29) to (3.35) for $k = 2$.

These equations can be written in the form

$$A^T \alpha + \alpha A + \beta(0) = -W \qquad (3.55)$$

$$\left(\alpha + \alpha^T\right)(A_1 + AB_1) + \beta(0)B_1 = 0 \qquad (3.56)$$

$$\left(\alpha + \alpha^T\right)(A_2 + AB_2) + \beta(0)B_2 - \beta(-r) = 0 \qquad (3.57)$$

$$A^T \beta(\theta) - \dfrac{d\beta(\theta)}{d\theta} + \delta^T(\theta, 0) = 0 \qquad (3.58)$$

$$(A_2 + AB_2)^T \beta(\theta) + B_2^T \delta^T(\theta, 0) - \delta(-r, \theta) = 0 \qquad (3.59)$$

$$(A_1 + AB_1)^T \beta(\theta) + B_1^T \delta^T(\theta, 0) = 0 \qquad (3.60)$$

43

$$\frac{\partial \delta(\theta, \sigma)}{\partial \theta} + \frac{\partial \delta(\theta, \sigma)}{\partial \sigma} = 0 \tag{3.61}$$

for $\theta \in [-r, 0]$, $\sigma \in [\theta, 0]$.

The solution of equation (3.61) is as below

$$\delta(\theta, \sigma) = \varphi(\theta - \sigma) \tag{3.62}$$

for $\theta \in [-r, 0]$, $\sigma \in [\theta, 0]$ where $\varphi \in C^1([-r, r], \mathbb{R}^{n \times n})$

C^1 is a space of continuous functions with continuous derivative.

It follows from equation (3.60) that

$$\delta^T(\theta, 0) = -\left(A + A_1 B_1^{-1}\right)^T \beta(\theta) \tag{3.63}$$

for $\theta \in [-r, 0]$

Putting the term (3.63) into formula (3.59) one attains

$$(A_2 - A_1 B_1^{-1} B_2)^T \beta(\theta) = \delta(-r, \theta) = \varphi(-r - \theta) \tag{3.64}$$

Substituting

$$\xi = -r - \theta \tag{3.65}$$

into (3.64) one obtains

$$\varphi(\xi) = (A_2 - A_1 B_1^{-1} B_2)^T \beta(-r - \xi) \tag{3.66}$$

for $\xi \in [-r, 0]$

Hence

$$\delta^T(\theta, 0) = \varphi^T(\theta) = \beta^T(-r - \theta)(A_2 - A_1 B_1^{-1} B_2) \tag{3.67}$$

for $\theta \in [-r, 0]$

Putting the relation (3.67) into (3.58) one attains

$$\frac{d\beta(\theta)}{d\theta} = A^T \beta(\theta) + \beta^T(-r - \theta)(A_2 - A_1 B_1^{-1} B_2) \tag{3.68}$$

for $\theta \in [-r, 0]$

The derivative of $\beta(-r - \theta)$ with respect to θ is computed

$$\frac{d\beta(-r - \theta)}{d\theta} = -\frac{d\beta(\xi)}{d\xi} =$$

$$= -A^T \beta(\xi) - \beta^T(-r - \xi)(A_2 - A_1 B_1^{-1} B_2) =$$

44

$$= -A^T \beta(-r - \theta) - \beta^T(\theta)(A_2 - A_1 B_1^{-1} B_2) \tag{3.69}$$

In such a manner one obtains a set of ordinary differential equations

$$\begin{cases} \frac{d\beta(\theta)}{d\theta} = A^T \beta(\theta) + \beta^T(-r - \theta)(A_2 - A_1 B_1^{-1} B_2) \\ \frac{d\beta(-r-\theta)}{d\theta} = -A^T \beta(-r - \theta) - \beta^T(\theta)(A_2 - A_1 B_1^{-1} B_2) \end{cases} \tag{3.70}$$

The new function is given

$$\kappa(\theta) = \beta(-r - \theta) \tag{3.71}$$

The set of equations (3.70) takes a form

$$\begin{cases} \frac{d\beta(\theta)}{d\theta} = A^T \beta(\theta) + \kappa^T(\theta)(A_2 - A_1 B_1^{-1} B_2) \\ \frac{d\kappa(\theta)}{d\theta} = -A^T \kappa(\theta) - \beta^T(\theta)(A_2 - A_1 B_1^{-1} B_2) \end{cases} \tag{3.72}$$

To obtain a solution of the equations (3.72) one needs the initial conditions $\beta(-r)$ and $\kappa(-r)$. The equation (3.71) implies that $\kappa(-r) = \beta(0)$ and therefore the equations (3.55) to (3.57) take a form

$$A^T \alpha + \alpha A + \kappa(-r) = -W \tag{3.73}$$

$$\left(\alpha + \alpha^T\right)(A_1 + AB_1) + \kappa(-r)B_1 = 0 \tag{3.74}$$

$$\left(\alpha + \alpha^T\right)(A_2 + AB_2) + \kappa(-r)B_2 - \beta(-r) = 0 \tag{3.75}$$

The set of algebraic equations (3.73) to (3.75) enables to obtain the matrix α and the initial conditions of the ordinary differential equations (3.72). The matrix α can be obtained from the relationship

$$\alpha^T A - A^T \alpha + \left(\alpha + \alpha^T\right) A_1 B_1^{-1} = W \tag{3.76}$$

Having the matrix α it is possible to receive

$$\beta(-r) = \left(\alpha + \alpha^T\right)(A_2 - A_1 B_1^{-1} B_2) \tag{3.77}$$

$$\kappa(-r) = -(\alpha + \alpha^T)(A_1 B_1^{-1} + A) \tag{3.78}$$

The matrix $\delta(\theta, \sigma)$ is given by the relation (3.62)

$$\delta(\theta, \sigma) = \varphi(\theta - \sigma) \tag{3.79}$$

for $\theta \in [-r,0]$, $\sigma \in [\theta,0]$. where φ is determined by the equation (3.66)

$$\varphi(\xi) = (A_2 - A_1 B_1^{-1} B_2)^T \beta(-r - \xi) =$$

$$= (A_2 - A_1 B_1^{-1} B_2)^T \kappa(\xi) \qquad (3.80)$$

hence

$$\delta(\theta, \sigma) = (A_2 - A_1 B_1^{-1} B_2)^T \kappa(\theta - \sigma) \qquad (3.81)$$

In this way one obtains all coefficients of the functional (3.25). This coefficients depend on the matrices A, A_i, B_i for $i = 1,2$ of the system (3.17). The time derivative of the functional (3.25) is negative definite. When the matrices α, $\beta(\theta)$ and $\delta(\theta, \sigma)$ for $\theta \in [-r,0]$, $\sigma \in [\theta,0]$ are positive definite the functional (3.25) becomes the Lyapunov functional.

3.4 An example. An inertial system with a delay and a PD controller.

Let us consider a first order inertial system with a delay described by an equation

$$\begin{cases} \frac{dx(t)}{dt} = -\frac{q}{T}x(t) + \frac{k_0}{T}u(t - r) \\ x(t_0) = x_o \\ x(t_0 + \theta) = 0 \\ u(t) = -kx(t) - T_d \frac{dx(t)}{dt} \end{cases} \qquad (3.82)$$

$t \geq t_0$, $x(t) \in \mathbb{R}$, $\theta \in [-r,0)$, k, k_0, T, T_d, q, $x_0 \in \mathbb{R}$, $r \geq 0$.

The parameter k_0 is a gain of a plant, k is a proportional gain, T_d is a derivative gain, T is a system time constant, x_0 is an initial state of a system. In the case $q = 1$ an equation (3.82) describes a static object and in the case $q = 0$ an equation (3.82) describes an astatic object.

One can reshape an equation (3.82) to a form

$$\begin{cases} \frac{dx(t)}{dt} + \frac{k_0 T_d}{T} \frac{dx(t-r)}{dt} = -\frac{q}{T}x(t) - \frac{k_0 k}{T}x(t - r) \\ x(t_0) = x_o \\ x(\theta) = 0 \end{cases} \qquad (3.83)$$

for $t \geq t_0$ and $\theta \in [-r,0)$.

It is assumed that the element $\frac{k_0 T_d}{T}$ satisfies a condition (3.15), whose takes a form

$$\left| \frac{k_0 T_d}{T} \right| < 1 \qquad (3.84)$$

A Lyapunov functional V is defined by a formula

$$V(y(t), x(t+\theta)) = \alpha y^2(t) + \int\limits_{-r}^{0} \beta(\theta) y(t) x(t+\theta) d\theta +$$

$$+ \int\limits_{-r}^{0} \int\limits_{\theta}^{0} \delta(\theta, \sigma) x(t+\theta) x(t+\sigma) d\sigma d\theta \qquad (3.85)$$

where

$$y(t) = x(t) + \frac{k_0 T_d}{T} x(t-r) \qquad (3.86)$$

In a parametric optimization problem is used an integral quadratic performance index of quality

$$J = \int\limits_{t_0}^{\infty} w y^2(t) dt = V(y(t_0), \Phi) \qquad (3.87)$$

The set of equations (3.50) takes a form

$$\begin{bmatrix} \frac{d\beta(\theta)}{d\theta} \\ \frac{d\kappa(\theta)}{d\theta} \end{bmatrix} = \begin{bmatrix} p_1 & -p_2 \\ p_2 & -p_1 \end{bmatrix} \begin{bmatrix} \beta(\theta) \\ \kappa(\theta) \end{bmatrix} \qquad (3.88)$$

where

$$p_1 = \frac{k_0^2 k T_d - q T}{T^2 - k_0^2 T_d^2} \qquad (3.89)$$

$$p_2 = \frac{k_0 k T - q k_0 T_d}{T^2 - k_0^2 T_d^2} \qquad (3.90)$$

The fundamental matrix of solutions of equation (3.88) is given by

$$R(\theta) = \begin{bmatrix} \cosh(\lambda\theta) + \frac{p_1}{\lambda}\sinh(\lambda\theta) & -\frac{p_2}{\lambda}\sinh(\lambda\theta) \\ \frac{p_2}{\lambda}\sinh(\lambda\theta) & \cosh(\lambda\theta) - \frac{p_1}{\lambda}\sinh(\lambda\theta) \end{bmatrix} \qquad (3.91)$$

where

$$\lambda = \sqrt{p_1^2 - p_2^2} = \sqrt{\frac{q^2 - k_0^2 k^2}{T^2 - k_0^2 T_d^2}} \qquad (3.92)$$

The solution of the set of equations (3.91) is given

$$\beta(\theta) = [\cosh(\lambda\theta + r) + \frac{p_1}{\lambda}\sinh(\lambda\theta + r)]\beta(-r) - \frac{p_2}{\lambda}\sinh(\lambda\theta + r)\kappa(-r) \qquad (3.93)$$

$$\kappa(\theta) = \frac{p_2}{\lambda}\sinh(\lambda\theta + r)\beta(-r) + [\cosh(\lambda\theta + r) - \frac{p_1}{\lambda}\sinh(\lambda\theta + r)]\kappa(-r) \qquad (3.94)$$

The initial conditions of equation (3.91) and the coefficient α are obtained from the set of algebraic equations (3.52) to (3.54) which take a form

$$[\cosh(\frac{\lambda r}{2}) + \frac{p_1 - p_2}{\lambda}\sinh(\frac{\lambda r}{2})]\beta(-r) + [-\cosh(\frac{\lambda r}{2}) + \frac{p_1 - p_2}{\lambda}\sinh(\frac{\lambda r}{2})]\kappa(-r) = 0 \qquad (3.95)$$

$$-\frac{2q}{T}\alpha + \kappa(-r) = -w \qquad (3.96)$$

$$2\frac{qk_0T_d - k_0kT}{T^2}\alpha - \frac{k_0T_d}{T}\kappa(-r) - \beta(-r) = 0 \qquad (3.97)$$

From equation (3.96) one can determine a term $\kappa(-r)$ and substitute it into (3.97) and (3.95).

$$\kappa(-r) = -w + \frac{2q}{T}\alpha \qquad (3.98)$$

From equation (3.97) one determines a term $\beta(-r)$

$$\beta(-r) = -2\frac{k_0k}{T}\alpha + \frac{k_0T_d}{T}w \qquad (3.99)$$

One puts the terms (3.98) and (3.99) into (3.95) and one obtains a parameter α

$$\alpha = \frac{w}{2}\frac{\frac{(T - k_0T_d)(p_1 - p_2)}{\lambda}\sinh(\frac{\lambda r}{2}) - (T + k_0T_d)\cosh(\frac{\lambda r}{2})}{\frac{(q - k_0k)(p_1 - p_2)}{\lambda}\sinh(\frac{\lambda r}{2}) - (q + k_0k)\cosh(\frac{\lambda r}{2})} \qquad (3.100)$$

The coefficient $\delta(\theta, \sigma)$ is obtained from equations (3.41), (3.42) and (3.88)

$$\delta(\theta, \sigma) = (p_1 + \frac{q}{T})\beta(\theta - \sigma) - p_2\kappa(\theta - \sigma) \qquad (3.101)$$

3.5   An example. A system with two delays and a PD controller

Let us consider a system described by an equation

$$\begin{cases} \frac{dx(t)}{dt} - b_1\frac{dx(t-\tau)}{dt} = ax(t) + a_1x(t - \tau) + k_0u(t - r) \\ u(t) = kx(t) + T_d\frac{dx(t)}{dt} \\ x(t_0 + \theta) = \Phi(\theta) \end{cases} \qquad (3.102)$$

$t \geq t_0$, $x(t) \in \mathbb{R}$, $\theta \in [-r, 0]$, $a, a_1, b_1, k_0, k, T_d \in \mathbb{R}$, $r \geq \tau > 0$

The parameter k_0 is a gain of a plant, k is a proportional gain, T_d is a derivative gain.

One can reshape an equation (3.102) to a form

$$\begin{cases} \frac{dx(t)}{dt} - b_1 \frac{dx(t-\tau)}{dt} - k_0 T_d \frac{dx(t-r)}{dt} = ax(t) + \\ \qquad\qquad\qquad +a_1 x(t-\tau) + k_0 kx(t-r) \\ x(t_0 + \theta) = \Phi(\theta) \end{cases} \qquad (3.103)$$

It is assumed that elements b_1 and $k_0 T_d$ satisfy the condition (3.15), whose takes a form

$$|b_1| + |k_0 T_d| < 1 \qquad (3.104)$$

The equation (3.103) is reshaped to a form

$$\begin{cases} \frac{dy(t)}{dt} = ay(t) + (a_1 + ab_1)x(t-\tau) + \\ \qquad\qquad + (k_0 k + ak_0 T_d)x(t-r) \\ y(t) = x(t) - b_1 x(t-\tau) - k_0 T_d x(t-r) \\ y(t_0) = \Phi(0) - b_1 \Phi(-\tau) - k_0 T_d \Phi(-r) \\ x(t_0 + \theta) = \Phi(\theta) \end{cases} \qquad (3.105)$$

A Lyapunov functional V is defined by a formula

$$V(y(t), x(t+\theta)) = \alpha y^2(t) + \int_{-r}^{0} y(t)\beta(\theta)x(t+\theta)d\theta +$$

$$+ \int_{-r}^{0}\int_{\theta}^{0} x(t+\theta)\delta(\theta,\sigma)x(t+\sigma)d\sigma d\theta \qquad (3.106)$$

In a parametric optimization problem is used an integral quadratic performance index of quality

$$J = \int_{t_0}^{\infty} wy^2(t)dt = V(y(t_0), \Phi) \qquad (3.107)$$

The set of equations (3.72) takes a form

$$\begin{bmatrix} \frac{d\beta(\theta)}{d\theta} \\ \frac{d\kappa(\theta)}{d\theta} \end{bmatrix} = \begin{bmatrix} a & \frac{k_0(b_1 k - a_1 T_d)}{b_1} \\ -\frac{k_0(b_1 k - a_1 T_d)}{b_1} & -a \end{bmatrix} \begin{bmatrix} \beta(\theta) \\ \kappa(\theta) \end{bmatrix} \qquad (3.108)$$

Fundamental matrix of solutions of the system (3.108) is given by a formula

$$R(\theta) = \begin{bmatrix} \cosh\lambda\theta + \frac{a}{\lambda}\sinh\lambda\theta & \frac{k_0(b_1 k - a_1 T_d)}{\lambda b_1}\sinh\lambda\theta \\ -\frac{k_0(b_1 k - a_1 T_d)}{\lambda b_1}\sinh\lambda\theta & \cosh\lambda\theta - \frac{a}{\lambda}\sinh\lambda\theta \end{bmatrix} \qquad (3.109)$$

where

$$\lambda = \sqrt{a^2 - \left(\frac{k_0(b_1 k - a_1 T_d)}{b_1}\right)^2} \qquad (3.110)$$

The solution of the equation (3.108) is given by a formula

$$\begin{bmatrix} \beta(\theta) \\ \kappa(\theta) \end{bmatrix} = R(\theta + r) \begin{bmatrix} \beta(-r) \\ \kappa(-r) \end{bmatrix} \qquad (3.111)$$

The parameter α and the initial conditions for the ordinary differential equation (3.108) are obtained from equations (3.76), (3.77) and (3.78)

$$\alpha = \frac{b_1 w}{2a_1} \qquad (3.112)$$

$$\beta(-r) = w\left(\frac{k_0 k b_1}{a_1} - k_0 T_d\right) \qquad (3.113)$$

$$\kappa(-r) = -w\left(1 + \frac{ab_1}{a_1}\right) \qquad (3.114)$$

Putting (3.113) and (3.114) into (3.111) one obtains

$$\beta(\theta) = w\frac{k_0(b_1 k - a_1 T_d)}{a_1}\cosh\lambda(\theta + r) - w\frac{k_0(b_1 k - a_1 T_d)}{\lambda b_1}\sinh\lambda(\theta + r) \qquad (3.115)$$

$$\kappa(\theta) = -w(1 + \frac{ab_1}{a_1})\cosh\lambda(\theta + r) + w(\frac{a}{\lambda} + \frac{b_1\lambda}{a_1})\sinh\lambda(\theta + r) \qquad (3.116)$$

It follows from (3.81) that the element δ is given by a formula

$$\delta(\theta, \sigma) = \left(k_0 k - \frac{a_1 k_0 T_d}{b_1}\right)\kappa(\theta - \sigma) \qquad (3.117)$$

Chapter 4

A linear system with a retarded type time-varying delay

4.1 A mathematical model of a linear system with a retarded type time-varying delay

Let us consider a linear system with a retarded type time-varying delay, whose dynamics is described by the equation

$$\begin{cases} \frac{dx(t)}{dt} = Ax(t) + Bx(t - \tau(t)) \\ x(t_0) = x_0 \in \mathbb{R}^n \\ x(t_0 + \theta) = \Phi(\theta) \end{cases} \tag{4.1}$$

where $A, B \in \mathbb{R}^{n \times n}$, $x(t) \in \mathbb{R}^n$, $\Phi \in W^{1,2}([-r,0), \mathbb{R}^n)$, $t \geq t_0$, $\theta \in [-r,0)$, $\tau(t)$ is a time-varying delay satisfying the condition $0 \leq \tau(t) \leq r$; $\frac{d\tau(t)}{dt} \neq 1$; $t \geq t_0$ where r is positive constant. $W^{1,2}([-r,0), \mathbb{R}^n)$ is a space of all absolutely continuous functions $[-r,0) \to \mathbb{R}^n$ with derivatives in $L^2([-r,0), \mathbb{R}^n)$ a space of Lebesgue square integrable functions on interval $[-r,0)$ with values in $\mathbb{R}^n$.

Definition 29. The norm in $W^{1,2}([-r,0), \mathbb{R}^n)$ is defined by

$$\| \Phi \|_{W^{1,2}}^2 = \int_{-r}^{0} \left(\| \Phi(t) \|_{\mathbb{R}^n}^2 + \| \frac{d\Phi(t)}{dt} \|_{\mathbb{R}^n}^2 \right) dt \tag{4.2}$$

where $\| \cdot \|_{\mathbb{R}^n}$ is an arbitrary norm in $\mathbb{R}^n$.

Definition 30. The function $x_t \in W^{1,2}([-r,0), \mathbb{R}^n)$ is called a shifted restriction of x to an interval $[t-r,t)$ and is defined by a formula

$$x_t(\theta) := x(t + \theta) \quad for \, t \geq t_0, \, \theta \in [-r,0) \tag{4.3}$$

Using the formula (4.3) one can write the equation (4.1) in the form

$$\begin{cases} \frac{dx(t)}{dt} = Ax(t) + Bx_t(-\tau(t)) \\ x(t_0) = x_0 \in \mathbb{R}^n \\ x_{t_0} = \Phi \in W^{1,2}([-r,0),\mathbb{R}^n) \end{cases} \qquad (4.4)$$

The state of the system (4.4) is a vector

$$S(t) = \begin{bmatrix} x(t) \\ x_t \end{bmatrix} \quad for\ t \geq t_0 \qquad (4.5)$$

The state space is defined by the formula

$$X = \mathbb{R}^n \times W^{1,2}([-r,0),\mathbb{R}^n) \qquad (4.6)$$

In a parametric optimization problem is used an integral quadratic performance index of quality

$$J = \int_{t_0}^{\infty} x^T(t)Wx(t)dt \qquad (4.7)$$

where $W \in \mathbb{R}^{n \times n}$ is a positive definite matrix.

4.2 A Lyapunov functional

Definition 31. A functional $V : X \to \mathbb{R}$ is **positive definite** if and only if it is continuous and $V(x) > 0$ for $x \neq 0$ and $V(0) = 0$.

A functional $V : X \to \mathbb{R}$ is **negative definite** if and only if it is continuous and $V(x) < 0$ for $x \neq 0$ and $V(0) = 0$.

A functional $V : X \times [t_0,\infty) \to \mathbb{R}$ is **positive definite** if it is continuous and there exists a positive definite functional $H : X \to \mathbb{R}$ such that $V(x,t) \geq H(x)$ and $V(0,t) = H(0) = 0$ for $x \in X$ and $t \geq t_0$.

Definition 32. A positive definite functional $V : X \times [t_0,\infty) \to \mathbb{R}$ is **upper bounded** if there exists a positive definite functional $H : X \to \mathbb{R}$ such that $V(x,t) \leq H(x)$ for $x \in X$ and $t \geq t_0$.

Definition 33. A time derivative of the functional $V(x(t),x_t,t)$ at $(x(t_0),\Phi,t_0)$ on a trajectory of a system (4.4) is given by the formula

$$\frac{dV(x(t_0),\Phi,t_0)}{dt} = \limsup_{h \to 0} \frac{1}{h} \left[V\left(x(t_0+h),x_{t_0+h},t_0+h\right) - V\left(x(t_0),\Phi,t_0\right) \right] \qquad (4.8)$$

Definition 34. The functional $V : X \times [t_0,\infty) \to \mathbb{R}$ is called a **Lyapunov functional** if

1. V is a positive definite upper bounded functional

2. V is differentiable

3. A time derivative of V computed according to a formula (4.8) on the trajectory of the system (4.4) is negative definite

Existence of the Lyapunov functional for the system (4.4) is a sufficient condition for asymptotic stability of its zero solution.

From the assumption that the Lyapunov functional is upper bounded results that there exists a functional H such that

$$0 \leq V(x(t),x_t,t) \leq H(x(t),x_t) \tag{4.9}$$

for $t \geq t_0$

When the system (4.4) is asymptotically stable

$$\lim_{t\to\infty} H(x(t),x_t) = 0 \text{ implies } \lim_{t\to\infty} V(x(t),x_t,t) = 0$$

Hence

$$\int_{t_0}^{\infty} \frac{dV(x(t),x_t,t)}{dt} dt = \lim_{t\to\infty} V(x(t),x_t,t) - \lim_{t\to t_0} V(x(t),x_t,t) =$$

$$= -V(\lim_{t\to t_0}(x(t),x_t,t)) = -V(x(t_0),\Phi,t_0) \tag{4.10}$$

Assume that the time derivative of the Lyapunov functional V is given as a quadratic form

$$\frac{dV(x(t),x_t,t)}{dt} \equiv -x^T(t)Wx(t) \tag{4.11}$$

for $t \geq t_0$, where $W \subset \mathbb{R}^{n\times n}$ is a positive definite matrix.

It follows from (4.7) and (4.11) that

$$J = \int_{t_0}^{\infty} x^T(t)Wx(t)dt = V(x(t_0),\Phi,t_0) \tag{4.12}$$

Corollary 35. *If one constructs a Lyapunov functional such that its time derivative computed on the trajectory of the system (4.4) is given as a quadratic form (4.11) one can not only investigate the system (4.4) stability but also calculate a value of a square indicator of quality (4.7) of the parametric optimization problem.*

To calculate the value of the performance index (4.7), which is equal to the value of the Lyapunov functional at the initial state of the system (4.4), one needs a mathematical formula of that functional.

4.3   Determination of the Lyapunov functional

Let us consider a quadratic functional on X, given by a formula

$$V(x(t),x_t,t) = x^T(t)\alpha(t)x(t) + \int_{-\tau(t)}^{0} x^T(t)\beta(\theta+\tau(t))x_t(\theta)d\theta +$$

$$+ \int\limits_{-\tau(t)}^{0} \int\limits_{\theta}^{0} x_t^T(\theta)\delta(\theta+\tau(t),\sigma+\tau(t))x_t(\sigma)d\sigma d\theta \qquad (4.13)$$

for $t \geq t_0$, where $\alpha \in C^1([t_0,\infty),\mathbb{R}^{n\times n})$, $\alpha(t)$ is positively defined $\beta \in C^1([0,\tau(t)],\mathbb{R}^{n\times n})$, $\delta \in C^1(\Omega,\mathbb{R}^{n\times n})$, $\Omega = \{(\theta,\sigma): \theta \in [0,\tau(t)], \sigma \in [\theta,0]\}, 0 \leq \tau(t) \leq r . C^1$ is a space of continuous functions with continuous derivative.

Conjecture 36. *It is given a procedure of determination of the functional (4.13) coefficients to obtain the Lyapunov functional.*

The time derivative of the functional (4.13) on the trajectory of the system (4.4) is computed. It is taken the following procedure. One computes the time derivative of each term of the right-hand-side of the formula (4.13) and one substitutes in place of $\frac{dx(t)}{dt}$ and $\frac{\partial x_t(\theta)}{\partial t}$ the following terms

$$\frac{dx(t)}{dt} = Ax(t) + Bx_t(-\tau(t)) \qquad (4.14)$$

$$\frac{\partial x_t(\theta)}{\partial t} = \frac{\partial x_t(\theta)}{\partial \theta} \qquad (4.15)$$

In such a manner one attains

$$\frac{dV(x(t),x_t,t)}{dt} = x^T(t)[A^T\alpha(t) + \alpha(t)A + \frac{d\alpha(t)}{dt} + \beta(\tau(t))]x(t)+$$

$$+x_t^T(-\tau(t))[B^T(\alpha(t)+\alpha^T(t)) + \beta^T(0)\left(\frac{d\tau(t)}{dt} - 1\right)]x(t)+$$

$$+ \int\limits_{-\tau(t)}^{0} x^T(t)[A^T\beta(\theta+\tau(t)) + \frac{d\beta(\theta+\tau(t))}{dt} - \frac{d\beta(\theta+\tau(t))}{d\theta}+$$

$$+\delta^T(\theta+\tau(t),\tau(t))]x_t(\theta)d\theta + \int\limits_{-\tau(t)}^{0} x_t^T(-\tau(t))[B^T\beta(\theta+\tau(t))+$$

$$+\delta(0,\theta+\tau(t))\left(\frac{d\tau(t)}{dt} - 1\right)]x_t(\theta)d\theta + \int\limits_{-\tau(t)}^{0}\int\limits_{\theta}^{0} x_t^T(\theta)[\frac{d\delta(\theta+\tau(t),\sigma+\tau(t))}{dt}+$$

$$-\frac{\partial\delta(\theta+\tau(t),\sigma+\tau(t))}{\partial\theta} - \frac{\partial\delta(\theta+\tau(t),\sigma+\tau(t))}{\partial\sigma}]x_t(\sigma)d\sigma d\theta \qquad (4.16)$$

for $t \geq t_0$ where $\alpha \in C^1([t_0,\infty),\mathbb{R}^{n\times n})$, $\beta \in C^1([0,\tau(t)],\mathbb{R}^{n\times n})$, $\delta \in C^1(\Omega,\mathbb{R}^{n\times n})$, $\Omega = \{(\theta,\sigma): \theta \in [0,\tau(t)],$ $0 \leq \tau(t) \leq r$.

One identifies the coefficients of the Lyapunov functional (4.13) assuming that the derivative (4.16) satisfies the relationship

$$\frac{dV(x(t),x_t,t)}{dt} \equiv -x^T(t)Wx(t) \tag{4.17}$$

for $t \geq t_0$, where $W \in \mathbb{R}^{n \times n}$ is positively defined matrix.

From equations (4.16) and (4.17) the set of equations is obtained

$$A^T \alpha(t) + \alpha(t)A + \frac{d\alpha(t)}{dt} + \beta(\tau(t)) = -W \tag{4.18}$$

$$B^T (\alpha(t) + \alpha^T(t)) + \beta^T(0)\left(\frac{d\tau(t)}{dt} - 1\right) = 0 \tag{4.19}$$

$$A^T \beta(\theta + \tau(t)) + \frac{d\beta(\theta + \tau(t))}{dt} - \frac{d\beta(\theta + \tau(t))}{d\theta} + \delta^T(\theta + \tau(t), \tau(t)) = 0 \tag{4.20}$$

$$B^T \beta(\theta + \tau(t)) + \delta(0, \theta + \tau(t))\left(\frac{d\tau(t)}{dt} - 1\right) = 0 \tag{4.21}$$

$$\frac{d\delta(\theta + \tau(t), \sigma + \tau(t))}{dt} - \frac{\partial\delta(\theta + \tau(t), \sigma + \tau(t))}{\partial\theta} - \frac{\partial\delta(\theta + \tau(t), \sigma + \tau(t))}{\partial\sigma} = 0 \tag{4.22}$$

for $t \geq t_0$; $\theta \in [-\tau(t), 0]$; $\sigma \in [\theta, 0]$ where $0 \leq \tau(t) \leq r$.

The new variables are introduced

$$\xi = \theta + \tau(t) \tag{4.23}$$

$$\eta = \sigma + \tau(t) \tag{4.24}$$

One calculates the derivatives

$$\frac{d\delta(\theta + \tau(t), \sigma + \tau(t))}{dt} = \frac{d\delta(\xi, \eta)}{dt} = \frac{\partial\delta(\xi, \eta)}{\partial\xi}\frac{d\tau(t)}{dt} + \frac{\partial\delta(\xi, \eta)}{\partial\eta}\frac{d\tau(t)}{dt} \tag{4.25}$$

$$\frac{\partial\delta(\theta + \tau(t), \sigma + \tau(t))}{\partial\theta} = \frac{\partial\delta(\xi, \eta)}{\partial\theta} = \frac{\partial\delta(\xi, \eta)}{\partial\xi} \tag{4.26}$$

$$\frac{\partial\delta(\theta + \tau(t), \sigma + \tau(t))}{\partial\sigma} = \frac{\partial\delta(\xi, \eta)}{\partial\sigma} = \frac{\partial\delta(\xi, \eta)}{\partial\eta} \tag{4.27}$$

$$\frac{d\beta(\theta + \tau(t))}{dt} = \frac{d\beta(\xi)}{d\xi}\frac{\partial\xi}{\partial t} = \frac{d\beta(\xi)}{d\xi}\frac{d\tau(t)}{dt} \tag{4.28}$$

$$\frac{d\beta(\theta + \tau(t))}{d\theta} = \frac{d\beta(\xi)}{d\xi}\frac{\partial\xi}{\partial\theta} = \frac{d\beta(\xi)}{d\xi} \tag{4.29}$$

The formula (4.22) takes the form

$$\frac{\partial \delta(\xi, \eta)}{\partial \xi} + \frac{\partial \delta(\xi, \eta)}{\partial \eta} = 0 \qquad (4.30)$$

for $t \geq t_0$, $\theta \in [-\tau(t), 0]$, $\sigma \in [\theta, 0]$, $\xi \in [0, \tau(t)]$, $\eta \in [\xi, \tau(t)]$ where $0 \leq \tau(t) \leq r$.
The solution of equation (4.22) is given by a formula

$$\delta(\theta + \tau(t), \sigma + \tau(t)) = \delta(\xi, \eta) = f(\xi - \eta) = f(\theta - \sigma) \qquad (4.31)$$

for $t \geq t_0$, $\theta \in [-\tau(t), 0]$, $\sigma \in [\theta, 0]$, $0 \leq \tau(t) \leq r$ where $f \in C^1([-r, r], \mathbb{R}^{n \times n})$
Taking into account a formula (4.31) one gets from equation (4.21) the relationship

$$\delta(0, \theta + \tau(t)) = f(-\tau(t) - \theta) = \left(1 - \frac{d\tau(t)}{dt}\right)^{-1} B^T \beta(\theta + \tau(t)) \qquad (4.32)$$

Hence

$$f(\xi) = \left(1 - \frac{d\tau(t)}{dt}\right)^{-1} B^T \beta(-\xi) \qquad (4.33)$$

for $\xi \in [0, \tau(t)]$ where $0 \leq \tau(t) \leq r$
The formula (4.31) implies

$$\delta^T(\theta + \tau(t), \tau(t)) = f^T(\theta) = \left(1 - \frac{d\tau(t)}{dt}\right)^{-1} \beta^T(-\theta) B \qquad (4.34)$$

After putting the term (4.34) into the formula (4.20) one obtains a relationship

$$A^T \beta(\theta + \tau(t)) + \frac{d\beta(\theta + \tau(t))}{dt} - \frac{d\beta(\theta + \tau(t))}{d\theta} + \left(1 - \frac{d\tau(t)}{dt}\right)^{-1} \beta^T(-\theta) B = 0 \qquad (4.35)$$

Taking into account the formulas (4.23), (4.28) and (4.29) one obtains from equation (4.35) the relationship

$$\frac{d\beta(\xi)}{d\xi} = -\left(\frac{d\tau(t)}{dt} - 1\right)^{-1} A^T \beta(\xi) + \left(\frac{d\tau(t)}{dt} - 1\right)^{-2} \beta^T(-\xi + \tau(t)) B \qquad (4.36)$$

for $\xi \in [0, \tau(t)]$ where $0 \leq \tau(t) \leq r$
Using the relationship (4.36) the derivative of the term $\beta(-\xi + \tau(t))$ with respect to ξ is calculated.
One obtains

$$\frac{d\beta(-\xi + \tau(t))}{d\xi} = -\left(\frac{d\tau(t)}{dt} - 1\right)^{-2} \beta^T(\xi) B + \left(\frac{d\tau(t)}{dt} - 1\right)^{-1} A^T \beta(-\xi + \tau(t)) \qquad (4.37)$$

for $\xi \in [0, \tau(t)]$ where $0 \leq \tau(t) \leq r$
In such a way one attains a set of differential equations

$$\begin{cases} \dfrac{d\beta(\xi)}{d\xi} = -\left(\dfrac{d\tau(t)}{dt}-1\right)^{-1}A^T\beta(\xi) + \\ \qquad +\left(\dfrac{d\tau(t)}{dt}-1\right)^{-2}\beta^T(-\xi+\tau(t))B \\ \dfrac{d\beta(-\xi+\tau(t))}{d\xi} = -\left(\dfrac{d\tau(t)}{dt}-1\right)^{-2}\beta^T(\xi)B+ \\ \qquad +\left(\dfrac{d\tau(t)}{dt}-1\right)^{-1}A^T\beta(-\xi+\tau(t)) \end{cases} \qquad (4.38)$$

for each fixed $t \geq t_0$, $\xi \in [0, \tau(t)]$ where $0 \leq \tau(t) \leq r$ with the initial conditions $\beta(0)$ and $\beta(\tau(t))$. There holds the relationship between $\beta(\xi)$ and $\beta(-\xi+\tau(t))$

$$\beta(\xi)\,|_{\xi=\frac{\tau(t)}{2}} = \beta(-\xi+\tau(t))\,|_{\xi=\frac{\tau(t)}{2}} \qquad (4.39)$$

The derivative of the equation (4.19) with respect to t is calculated

$$B^T\left(\frac{d\alpha(t)}{dt}+\frac{d\alpha^T(t)}{dt}\right)+\frac{d\beta^T(0)}{dt}\left(\frac{d\tau(t)}{dt}-1\right)+\beta^T(0)\frac{d^2\tau(t)}{dt^2}=0 \qquad (4.40)$$

From equation (4.36) it results that

$$\frac{d\beta^T(0)}{dt}=-\frac{d\tau(t)}{dt}\left(\frac{d\tau(t)}{dt}-1\right)^{-1}\beta^T(0)A+\frac{d\tau(t)}{dt}\left(\frac{d\tau(t)}{dt}-1\right)^{-2}B^T\beta(\tau(t)) \qquad (4.41)$$

The equation (4.18) implies

$$\frac{d\alpha(t)}{dt}=-A^T\alpha(t)-\alpha(t)A-\beta(\tau(t))-W \qquad (4.42)$$

One puts the terms (4.41) and (4.42) into equation (4.40). After calculations one attains

$$B^T\left[A^T\left(\alpha(t)+\alpha^T(t)\right)+\left(\alpha(t)+\alpha^T(t)\right)A\right]+\beta^T(0)\left(\frac{d\tau(t)}{dt}A-\frac{d^2\tau(t)}{dt^2}I\right)+$$

$$-\left(\frac{d\tau(t)}{dt}-1\right)^{-1}B^T\beta(\tau(t))+B^T\beta^T(\tau(t))=-B^T\left(W+W^T\right) \qquad (4.43)$$

Solving the set of equations (4.43), (4.19) and (4.39) one obtains the matrix $\alpha(t)$ and the initial conditions of the system (4.38). That set of equations is written below

$$B^T\left[A^T\left(\alpha(t)+\alpha^T(t)\right)+\left(\alpha(t)+\alpha^T(t)\right)A\right]+\beta^T(0)\left(\frac{d\tau(t)}{dt}A-\frac{d^2\tau(t)}{dt^2}I\right)+$$

$$-\left(\frac{d\tau(t)}{dt}-1\right)^{-1}B^T\beta(\tau(t))+B^T\beta^T(\tau(t))=-B^T\left(W+W^T\right) \qquad (4.44)$$

$$B^T\left(\alpha(t)+\alpha^T(t)\right)+\beta^T(0)\left(\frac{d\tau(t)}{dt}-1\right)=0 \qquad (4.45)$$

$$\beta(\xi)\,|_{\xi=\frac{\tau(t)}{2}} = \beta(-\xi+\tau(t))\,|_{\xi=\frac{\tau(t)}{2}} \qquad (4.46)$$

Having the solution of the set of differential equations (4.38) and taking into account the formulas (4.23), (4.31) and (4.33) one can get the matrices

$$\beta\left(\theta+\tau(t)\right)=\beta\left(\xi\right)|_{\xi=\theta+\tau(t)} \tag{4.47}$$

$$\delta\left(\theta+\tau(t),\sigma+\tau(t)\right)=\left(1-\frac{d\tau(t)}{dt}\right)^{-1}B^{T}\beta\left(\sigma-\theta\right) \tag{4.48}$$

for $t\geq t_{0}$, $\theta\in[-\tau(t),0]$, $\sigma\in[\theta,0]$ where $0\leq\tau(t)\leq r$.

In this way one obtained all coefficients of the functional (4.13). This coefficients depend on the matrices A and B of the system (4.4). The time derivative of the functional (4.13) is negative definite. When the matrices $\alpha(t)$, $\beta(\theta+\tau(t))$ and $\delta(\theta+\tau(t),\sigma+\tau(t))$ for $t\geq t_{0}, \theta\in[-\tau(t),0]$, $\sigma\in[\theta,0]$ are positive definite then the functional (4.13) becomes the Lyapunov functional.

4.4 The example. An inertial system with a delay and a P controller.

Let us consider a first order inertial system with a delay described by an equation

$$\begin{cases} \frac{dx(t)}{dt}=-\frac{q}{T}x(t)+\frac{k_0}{T}u(t-\tau(t)) \\ x(t_0)=x_o \\ x(t_0+\theta)=\Phi(\theta) \\ u(t)=-kx(t) \end{cases} \tag{4.49}$$

$t\geq t_0, x(t)\in\mathbb{R}, \Phi\in W^{1,2}([-r,0),\mathbb{R}), \theta\in[-r,0), k, k_0, T, q, x_0\in\mathbb{R}, r\geq 0, \tau(t)$ is a time-varying delay satisfying the condition $0\leq\tau(t)\leq r; \frac{d\tau(t)}{dt}\neq 1; t\geq t_0$ where r is positive constant. The parameter k_0 is a gain of a plant, k is a gain of a P controller, T is a system time constant, x_0 is an initial state of a system. In the case $q=1$ an equation (4.49) describes a static object and in the case $q=0$ an equation (4.49) describes an astatic object.

One can reshape an equation (4.49) to a form

$$\begin{cases} \frac{dx(t)}{dt}=-\frac{q}{T}x(t)-\frac{k_0 k}{T}x(t-\tau(t)) \\ x(t_0)=x_0 \\ x(t_0+\theta)=\Phi(\theta) \end{cases} \tag{4.50}$$

Definition 37. The function $x_t\in W^{1,2}([-r,0),\mathbb{R}^n)$ is called a shifted restriction of x to an interval $[t-r,t)$ and is defined by a formula

$$x_t(\theta):=x(t+\theta) \tag{4.51}$$

for $t\geq t_0, \theta\in[-r,0)$.

Using the formula (4.51) one can write the equation (4.50) in the form

$$\begin{cases} \frac{dx(t)}{dt} = -\frac{q}{T}x(t) - \frac{k_0 k}{T}x_t\left(-\tau\left(t\right)\right) \\ x\left(t_0\right) = x_0 \\ x_{t_0} = \Phi \end{cases} \qquad (4.52)$$

The Lyapunov functional is given by a formula

$$V\left(x(t),x_t,t\right) = \alpha(t)x^2\left(t\right) + \int\limits_{-\tau(t)}^{0} \beta\left(\theta + \tau(t)\right)x(t)x_t\left(\theta\right)d\theta +$$

$$+ \int\limits_{-\tau(t)}^{0} \int\limits_{\theta}^{0} \delta\left(\theta + \tau(t), \sigma + \tau(t)\right)x_t\left(\theta\right)x_t\left(\sigma\right)d\sigma d\theta \qquad (4.53)$$

The coefficients of the functional (4.53) will be obtained.

The equation (4.38) takes the form

$$\begin{bmatrix} \frac{d\beta(\xi)}{d\theta} \\ \frac{d\beta(-\xi+\tau(t))}{d\theta} \end{bmatrix} = \begin{bmatrix} \frac{-q}{T\left(1-\frac{d\tau(t)}{dt}\right)} & \frac{-k_0 k}{T\left(1-\frac{d\tau(t)}{dt}\right)^2} \\ \frac{k_0 k}{T\left(1-\frac{d\tau(t)}{dt}\right)^2} & \frac{q}{T\left(1-\frac{d\tau(t)}{dt}\right)} \end{bmatrix} \begin{bmatrix} \beta(\xi) \\ \beta\left(-\xi+\tau(t)\right) \end{bmatrix} \qquad (4.54)$$

for $t \geq t_0$, $\xi \in [0, \tau(t)]$ where $0 \leq \tau(t) \leq r$.

The fundamental matrix of the differential equation (4.54) is given by formula

$$R(\xi) = \begin{bmatrix} ch\lambda\xi - \frac{q}{\lambda T\left(1-\frac{d\tau(t)}{dt}\right)}sh\lambda\xi & -\frac{k_0 k}{\lambda T\left(1-\frac{d\tau(t)}{dt}\right)^2}sh\lambda\xi \\ \frac{k_0 k}{\lambda T\left(1-\frac{d\tau(t)}{dt}\right)^2}sh\lambda\xi & ch\lambda\xi + \frac{q}{\lambda T\left(1-\frac{d\tau(t)}{dt}\right)}sh\lambda\xi \end{bmatrix} \qquad (4.55)$$

where

$$\lambda = \frac{\sqrt{q^2\left(1 - \frac{d\tau(t)}{dt}\right)^2 - k_0^2 k^2}}{T\left(1 - \frac{d\tau(t)}{dt}\right)^2} \qquad (4.56)$$

Hence

$$\begin{bmatrix} \beta(\xi) \\ \beta\left(-\xi+\tau(t)\right) \end{bmatrix} = R(\xi)\begin{bmatrix} \beta(0) \\ \beta(\tau(t)) \end{bmatrix} \qquad (4.57)$$

for $t \geq t_0$, $\xi \in [0, \tau(t)]$ where $0 \leq \tau(t) \leq r$.

One needs the initial conditions of the set of differential equations (4.54) to obtain

$$\beta\left(\theta + \tau(t)\right) = \beta\left(\xi\right)|_{\xi = \theta + \tau(t)} \qquad (4.58)$$

$$\delta\left(\theta+\tau(t),\sigma+\tau(t)\right)=-\frac{k_0k}{T}\left(1-\frac{d\tau(t)}{dt}\right)^{-1}\beta\left(\sigma-\theta\right) \qquad (4.59)$$

for $t\geq t_0$, $\theta\in[-\tau(t),0]$, $\sigma\in[\theta,0]$ where $0\leq\tau(t)\leq r$.

The initial conditions of the differential equation (4.54) and the coefficient $\alpha(t)$ are obtained by solving of a set of equations (4.44) to (4.46) which take the form as below

$$4\frac{qk_0k}{T^2}\alpha(t)+\left(-\frac{q}{T}\frac{d\tau(t)}{dt}-\frac{d^2\tau(t)}{dt^2}\right)\beta(0)+\left(1-\frac{1}{\frac{d\tau(t)}{dt}-1}\right)b\beta(\tau(t))=-2bw \qquad (4.60)$$

$$-\frac{2k_0k}{T}\alpha(t)+\left(\frac{d\tau(t)}{dt}-1\right)\beta(0)=0 \qquad (4.61)$$

$$p_1\beta(0)+p_2\beta(\tau(t))=0 \qquad (4.62)$$

where

$$p_1=ch\frac{\lambda\tau(t)}{2}+\left(-\frac{q}{\lambda T\left(1-\frac{d\tau(t)}{dt}\right)}-\frac{k_0k}{\lambda T\left(1-\frac{d\tau(t)}{dt}\right)^2}\right)sh\frac{\lambda\tau(t)}{2} \qquad (4.63)$$

$$p_2=-ch\frac{\lambda\tau(t)}{2}+\left(-\frac{q}{\lambda T\left(1-\frac{d\tau(t)}{dt}\right)}-\frac{k_0k}{\lambda T\left(1-\frac{d\tau(t)}{dt}\right)^2}\right)sh\frac{\lambda\tau(t)}{2} \qquad (4.64)$$

Chapter 5

A linear neutral system with a time-varying delay

5.1 A mathematical model of a linear neutral system with a time-varying delay

Let us consider a linear neutral system with a time-varying delay, whose dynamics is described by the functional-differential equation (FDE)

$$
\begin{cases}
\frac{dx(t)}{dt} - C\frac{dx(t-\tau(t))}{dt} = Ax(t) + Bx(t - \tau(t)) \\
x(t_0) = x_0 \in \mathbb{R}^n \\
x(t_0 + \theta) = \Phi(\theta)
\end{cases}
\tag{5.1}
$$

where $t \geq t_0$, $\theta \in [-r, 0)$, $\tau(t)$ is a time-varying delay satisfying the condition $0 \leq \tau(t) \leq r$, $\frac{d\tau(t)}{dt} \neq 1$ where r is a positive constant A, B, $C \in \mathbb{R}^{n \times n}$ and C is non-singular, $x(t) \in \mathbb{R}^n$, $\Phi \in W^{1,2}([-r, 0), \mathbb{R}^n)$ $W^{1,2}([-r, 0), \mathbb{R}^n)$ is a space of all absolutely continuous functions $[-r, 0) \to \mathbb{R}^n$ with derivatives in $L^2([-r, 0), \mathbb{R}^n)$ a space of Lebesgue square integrable functions on an interval $[-r, 0)$ with values in $\mathbb{R}^n$.

Definition 38. The norm in $W^{1,2}([-r, 0), \mathbb{R}^n)$ is defined by

$$
\| \Phi \|_{W^{1,2}}^2 = \int_{-r}^{0} \left(\| \Phi(t) \|_{\mathbb{R}^n}^2 + \| \frac{d\Phi(t)}{dt} \|_{\mathbb{R}^n}^2 \right) dt
\tag{5.2}
$$

where $\| \cdot \|_{\mathbb{R}^n}$ is an arbitrary norm in $\mathbb{R}^n$.

The space of initial data is given by the Cartesian product $\mathbb{R}^n \times W^{1,2}([-r, 0), \mathbb{R}^n)$.

One can obtain a solution of FDE (5.1) using a step method. The step method is a basic method for solving FDE with a lumped delay. A solution is found on successive intervals, one after another, by solving an ordinary equation without delay in each interval.

A solution of the equation (5.1) is an absolutely continuous function defined for $t \geq t_0 - r$ with values in $\mathbb{R}^n$.

$$x(\cdot) \in W^{1,2}([t_0 - r, \infty), \mathbb{R}^n) \tag{5.3}$$

Where $W^{1,2}([t_0 - r, \infty), \mathbb{R}^n)$ is a space of all absolutely continuous functions with derivatives in a space of Lebesgue square integrable functions on interval $[t_0 - r, \infty)$ with values in $\mathbb{R}^n$.

Definition 39. The zero solution of (5.1) is **stable** if for any $\varepsilon > 0$ there is a $\delta > 0$ such that

$$\sqrt{\| x(t_0) \|_{\mathbb{R}^n}^2 + \| \Phi \|_{W^{1,2}}^2} < \delta$$

implies $\| x(t) \|_{\mathbb{R}^n} \leq \varepsilon$ for $t \geq t_0$.

Definition 40. The zero solution of (5.1) is **asymptotically stable** if

$$\| x(t) \|_{\mathbb{R}^n} \to 0$$

as $t \to \infty$.

Definition 41. The difference equation associated with (5.1) is given by

$$x(t) = Cx(t - \tau(t)) \tag{5.4}$$

for $t \geq t_0$.

The eigenvalues of the difference equation (5.4) play a fundamental role in the asymptotic behavior of the solutions of the neutral equation (5.1). The difference equation (5.4) is stable when the spectral radius $\gamma(C)$ of the matrix C fulfills the condition

$$\gamma(C) = sup\{| \lambda |: \lambda \in \sigma(C)\} < 1 \tag{5.5}$$

where the spectrum $\sigma(C)$ is the set of complex numbers λ for which the matrix $\lambda I - C$ is not invertible.

A new function y is introduced and defined by a term

$$y(t) = x(t) - Cx(t - \tau(t)) \tag{5.6}$$

for $t \geq t_0$.

Thus the equation (5.1) takes a form

$$\begin{cases} \frac{dy(t)}{dt} = Ay(t) + (AC + B)x(t - \tau(t)) \\ y(t) = x(t) - Cx(t - \tau(t)) \\ y(t_0) = x_0 - C\Phi(-\tau(t)) \\ x(t_0 + \theta) = \Phi(\theta) \end{cases} \tag{5.7}$$

It is assumed that the matrix C fulfills the condition (5.5)

Definition 42. The function $x_t \in W^{1,2}([-r,0),\mathbb{R}^n)$ is called a shifted restriction of x to an interval $[t-r,t)$ and is defined by a formula

$$x_t(\theta) := x(t+\theta) \tag{5.8}$$

for $t \geq t_0$, $\theta \in [-r,0)$.

Using the formula (5.8) one can write the equation (4.1) in the form

$$\begin{cases} \frac{dy(t)}{dt} = Ay(t) + (AC+B)x_t(-\tau(t)) \\ y(t) = x(t) - Cx(t-\tau(t)) \\ y(t_0) = x_0 - C\Phi(-\tau(t)) \\ x_{t_0} = \Phi \end{cases} \tag{5.9}$$

The state of the system (5.7) is a vector

$$S(t) = \begin{bmatrix} y(t) \\ x_t \end{bmatrix} \tag{5.10}$$

for $t \geq t_0$, where $y(t) \in \mathbb{R}^n$, $x_t \in W^{1,2}([-r,0),\mathbb{R}^n)$ and $x_t(\theta) = x(t+\theta)$ for $\theta \in [-r,0)$. The state space is defined by the formula

$$X = \mathbb{R}^n \times W^{1,2}([-r,0),\mathbb{R}^n) \tag{5.11}$$

The norm in the state space X is defined by

$$\| S(t) \|_X^2 = \| y(t) \|_{\mathbb{R}^n}^2 + \| x_t \|_{W^{1,2}}^2 \tag{5.12}$$

for $t \geq t_0$.

In a parametric optimization problem is used an integral quadratic performance index of quality

$$J = \int_{t_0}^{\infty} y^T(t)Wy(t)dt \tag{5.13}$$

where $W \in \mathbb{R}^{n \times n}$ is a positive definite matrix

5.2 A Lyapunov functional

Definition 43. A functional $V : X \to \mathbb{R}$ is **positive definite** if and only if it is continuous and $V(x) > 0$ for $x \neq 0$ and $V(0) = 0$.

A functional $V : X \to \mathbb{R}$ is **negative definite** if and only if it is continuous and $V(x) < 0$ for $x \neq 0$ and $V(0) = 0$.

A functional $V : X \times [t_0, \infty) \to \mathbb{R}$ is **positive definite** if it is continuous and there exists a positive definite functional $H : X \to \mathbb{R}$ such that $V(x,t) \geq H(x)$ and $V(0,t) = H(0) = 0$ for $x \in X$ and $t \geq t_0$.

Definition 44. A positive definite functional $V : X \times [t_0, \infty) \to \mathbb{R}$ is **upper bounded** if there exists a positive definite functional $H : X \to \mathbb{R}$ such that $V(x,t) \leq H(x)$ for $x \in X$ and $t \geq t_0$.

Definition 45. The time derivative of the functional $V(y(t), x_t, t)$ at $(y(t_0), \Phi, t_0)$ on a trajectory of a system (5.7) is defined by the formula

$$\frac{dV(y(t_0), \Phi, t_0)}{dt} = \limsup_{h \to 0} \frac{1}{h} \left[V\left(y(t_0 + h), x_{t_0+h}, t_0 + h\right) - V\left(y(t_0), \Phi, t_0\right) \right] \tag{5.14}$$

Definition 46. $V : X \times [t_0, \infty) \to \mathbb{R}$ is said a **Lyapunov functional** if

1. V is a positive definite upper bounded functional

2. V is differentiable

3. A time derivative of V computed according to a formula (5.14) on the trajectory of the system (5.7) is negative definite

Existence of the Lyapunov functional for the system (5.7) is a sufficient condition for asymptotic stability of its zero solution.

The assumption that the Lyapunov functional is upper bounded implies that there exists a functional H such that

$$0 \leq V(y(t), x_t, t) \leq H(y(t), x_t) \tag{5.15}$$

for $t \geq t_0$

When the system (5.7) is asymptotically stable

$$\lim_{t \to \infty} H(y(t), x_t) = 0 \; implies \; \lim_{t \to \infty} V(y(t), x_t, t) = 0$$

Hence

$$\int_{t_0}^{\infty} \frac{dV(y(t), x_t, t)}{dt} dt = \lim_{t \to \infty} V(y(t), x_t, t) - \lim_{t \to t_0} V(y(t), x_t, t) =$$

$$= -V\left(\lim_{t \to t_0} (y(t), x_t, t)\right) = -V(y(t_0), \Phi, t_0) \tag{5.16}$$

One assumes that the time derivative of the Lyapunov functional V is given as a quadratic form

$$\frac{dV(y(t), x_t, t)}{dt} \equiv -y^T(t) W y(t) \tag{5.17}$$

for $t \geq t_0$ where $W \in \mathbb{R}^{n \times n}$ is a positive definite matrix.

Taking (5.16) and (5.17) into account one obtains a relationship

$$J = \int_{t_0}^{\infty} y^T(t) W y(t) dt = V(y_0, \Phi, t_0) \tag{5.18}$$

Corollary 47. *If we construct a Lyapunov functional such that its time derivative computed on the trajectory of the system (5.7) will be given as a quadratic form (5.17) we can not only investigate the system (5.7) stability but also we can calculate a value of a square indicator of quality (5.18) of the parametric optimization problem.*

To calculate the value of the performance index (5.18), which is equal to the value of the Lyapunov functional at the initial state of the system (5.7), one needs a mathematical formula of the functional.

5.3   Determination of the Lyapunov functional

Let us consider a quadratic functional on $X \times [t_0, \infty)$, where X is defined by (5.11), given by a formula

$$V(y(t), x_t, t) = y^T(t)\alpha(t)y(t) + \int_{-\tau(t)}^{0} y^T(t)\beta(\theta + \tau(t))x_t(\theta)d\theta +$$

$$+ \int_{-\tau(t)}^{0} \int_{\theta}^{0} x_t^T(\theta)\delta(\theta + \tau(t), \sigma + \tau(t))x_t(\sigma)d\sigma d\theta \tag{5.19}$$

for $t \geq t_0$ where $\alpha \in C^1([t_0, \infty), \mathbb{R}^{n \times n})$, $\beta \in C^1([0, \tau(t)], \mathbb{R}^{n \times n})$, $\delta \in C^1(\Omega, \mathbb{R}^{n \times n})$, $\Omega = \{(\theta, \sigma) : \theta \in [0, \tau(t)], \sigma \in [\theta, 0 \leq \tau(t) \leq r$.

where C^1 is a space of all continuous functions with continuous derivative.

Conjecture 48. *We introduce a procedure of determination of the functional (5.19) coefficients to obtain the Lyapunov functional*

The time derivative of the functional (5.19) on the trajectory of the system (5.9) is computed. This time derivative is defined by the formula (5.14). It is taken the following procedure. One computes the time derivative of each term of the right-hand-side of the formula (5.19) and one substitutes in place of $\frac{dy(t)}{dt}$ and $\frac{\partial x_t(\theta)}{\partial t}$ the following terms

$$\frac{dy(t)}{dt} = Ay(t) + (AC + B)x_t(-\tau(t)) \tag{5.20}$$

$$\frac{\partial x_t(\theta)}{\partial t} = \frac{\partial x_t(\theta)}{\partial \theta} \tag{5.21}$$

In such a manner one attains

$$\frac{dV(y(t), x_t, t)}{dt} = y^T(t)[A^T\alpha(t) + \alpha(t)A + \frac{d\alpha(t)}{dt} + \beta(\tau(t))]y(t) +$$

$$+ y^T(t)[(\alpha(t) + \alpha^T(t))(AC + B) + \beta(\tau(t))C + \beta(0)\left(\frac{d\tau(t)}{dt} - 1\right)]x_t(-\tau(t)) +$$

$$+ \int_{-\tau(t)}^{0} y^T(t)[A^T \beta(\theta + \tau(t)) + \frac{d\beta(\theta + \tau(t))}{dt} - \frac{d\beta(\theta + \tau(t))}{d\theta} +$$

$$+ \delta^T(\theta + \tau(t), \tau(t))]x_t(\theta)d\theta + \int_{-\tau(t)}^{0} x_t^T(-\tau(t))[(AC+B)^T \beta(\theta + \tau(t)) +$$

$$+ C^T \delta^T(\theta + \tau(t), \tau(t)) + \delta(0, \theta + \tau(t)) \left(\frac{d\tau(t)}{dt} - 1 \right)]x_t(\theta)d\theta +$$

$$+ \int_{-\tau(t)}^{0} \int_{\theta}^{0} x_t^T(\theta)[\frac{d\delta(\theta + \tau(t), \sigma + \tau(t))}{dt} - \frac{\partial \delta(\theta + \tau(t), \sigma + \tau(t))}{\partial \theta} +$$

$$- \frac{\partial \delta(\theta + \tau(t), \sigma + \tau(t))}{\partial \sigma}]x_t(\sigma)d\sigma d\theta \qquad (5.22)$$

for $t \geq t_0$ where $\alpha \in C^1([t_0, \infty), \mathbb{R}^{n \times n})$, $\beta \in C^1([0, \tau(t)], \mathbb{R}^{n \times n})$, $\delta \in C^1(\Omega, \mathbb{R}^{n \times n})$, $\Omega = \{(\theta, \sigma) : \theta \in [0, \tau(t)], \sigma \in [\theta, 0]\}$, $0 \leq \tau(t) \leq r$.

The time derivative of the Lyapunov functional should be negative definite, therefore one identifies the coefficients of the functional (5.19) assuming that the time derivative (5.22) satisfies the relationship (5.17).

From equations (5.22) and (5.17) one obtains the set of equations

$$A^T \alpha(t) + \alpha(t)A + \frac{d\alpha(t)}{dt} + \beta(\tau(t)) = -W \qquad (5.23)$$

$$\left(\alpha(t) + \alpha^T(t) \right)(AC+B) + \beta(\tau(t))C + \beta(0) \left(\frac{d\tau(t)}{dt} - 1 \right) = 0 \qquad (5.24)$$

$$A^T \beta(\theta + \tau(t)) + \frac{d\beta(\theta + \tau(t))}{dt} - \frac{d\beta(\theta + \tau(t))}{d\theta} + \delta^T(\theta + \tau(t), \tau(t)) = 0 \qquad (5.25)$$

$$(AC+B)^T \beta(\theta + \tau(t)) + C^T \delta^T(\theta + \tau(t), \tau(t)) + \delta(0, \theta + \tau(t)) \left(\frac{d\tau(t)}{dt} - 1 \right) = 0 \qquad (5.26)$$

$$\frac{d\delta(\theta + \tau(t), \sigma + \tau(t))}{dt} - \frac{\partial \delta(\theta + \tau(t), \sigma + \tau(t))}{\partial \theta} - \frac{\partial \delta(\theta + \tau(t), \sigma + \tau(t))}{\partial \sigma} = 0 \qquad (5.27)$$

for $t \geq t_0$; $\theta \in [-\tau(t), 0]$; $\sigma \in [\theta, 0]$ where $0 \leq \tau(t) \leq r$.

The new variables are introduced

$$\xi = \theta + \tau(t) \qquad (5.28)$$

$$\eta = \sigma + \tau(t) \qquad (5.29)$$

The derivatives are calculated

$$\frac{d\delta\left(\theta+\tau(t),\sigma+\tau(t)\right)}{dt} = \frac{d\delta(\xi,\eta)}{dt} = \frac{\partial\delta(\xi,\eta)}{\partial\xi}\frac{d\tau(t)}{dt} + \frac{\partial\delta(\xi,\eta)}{\partial\eta}\frac{d\tau(t)}{dt} \tag{5.30}$$

$$\frac{\partial\delta\left(\theta+\tau(t),\sigma+\tau(t)\right)}{\partial\theta} = \frac{\partial\delta(\xi,\eta)}{\partial\theta} = \frac{\partial\delta(\xi,\eta)}{\partial\xi} \tag{5.31}$$

$$\frac{\partial\delta\left(\theta+\tau(t),\sigma+\tau(t)\right)}{\partial\sigma} = \frac{\partial\delta(\xi,\eta)}{\partial\sigma} = \frac{\partial\delta(\xi,\eta)}{\partial\eta} \tag{5.32}$$

$$\frac{d\beta\left(\theta+\tau(t)\right)}{dt} = \frac{d\beta(\xi)}{d\xi}\frac{\partial\xi}{\partial t} = \frac{d\beta(\xi)}{d\xi}\frac{d\tau(t)}{dt} \tag{5.33}$$

$$\frac{d\beta\left(\theta+\tau(t)\right)}{d\theta} = \frac{d\beta(\xi)}{d\xi}\frac{\partial\xi}{\partial\theta} = \frac{d\beta(\xi)}{d\xi} \tag{5.34}$$

The formula (5.27) takes a form

$$\frac{\partial\delta(\xi,\eta)}{\partial\xi} + \frac{\partial\delta(\xi,\eta)}{\partial\eta} = 0 \tag{5.35}$$

for $t \geq t_0$, $\theta \in [-\tau(t),0]$, $\sigma \in [\theta,0]$, $\xi \in [0,\tau(t)]$, $\eta \in [\xi,\tau(t)]$ where $0 \leq \tau(t) \leq r$.
The formula (5.25) takes a form

$$\left(\frac{d\tau(t)}{dt} - 1\right)\frac{d\beta(\xi)}{d\xi} + A^T\beta(\xi) + \delta^T\left(\xi,\tau(t)\right) = 0 \tag{5.36}$$

The formula (5.26) takes a form

$$(AC+B)^T\beta(\xi) + C^T\delta^T\left(\xi,\tau(t)\right) + \delta\left(0,\xi\right)\left(\frac{d\tau(t)}{dt} - 1\right) = 0 \tag{5.37}$$

The solution of the equation (5.27) is given by a formula

$$\delta\left(\theta+\tau(t),\sigma+\tau(t)\right) = \delta\left(\xi,\eta\right) = f(\xi-\eta) = f\left(\theta-\sigma\right) \tag{5.38}$$

for $t \geq t_0$, $\theta \in [-\tau(t),0]$, $\sigma \in [\theta,0]$, $0 \leq \tau(t) \leq r$ where $f \in C^1\left([-r,r],\mathbb{R}^{n\times n}\right)$
The formula (5.36) implies

$$\delta^T\left(\xi,\tau(t)\right) = f^T\left(\xi-\tau(t)\right) = -\left(\frac{d\tau(t)}{dt} - 1\right)\frac{d\beta(\xi)}{d\xi} - A^T\beta(\xi) \tag{5.39}$$

One puts the term (5.39) into (5.37). After calculations one obtains

$$C^T\frac{d\beta(\xi)}{d\xi} = \left(\frac{d\tau(t)}{dt} - 1\right)^{-1}B^T\beta(\xi) + \delta\left(0,\xi\right) \tag{5.40}$$

From the relation (5.39) one can determine the term $\delta(0,\xi) = f(-\xi)$

$$f(-\xi) = \left(\frac{d\tau(t)}{dt} - 1\right)\frac{d\beta^T(-\xi+\tau(t))}{d\xi} - \beta^T(-\xi+\tau(t))A \qquad (5.41)$$

and put it into (5.40). In this way the relation is obtained

$$C^T\frac{d\beta(\xi)}{d\xi} - \left(\frac{d\tau(t)}{dt} - 1\right)\frac{d\beta^T(-\xi+\tau(t))}{d\xi} = \left(\frac{d\tau(t)}{dt} - 1\right)^{-1}B^T\beta(\xi) - \beta^T(-\xi+\tau(t))A \quad (5.42)$$

for $\xi \in [0, \tau(t)]$ where $0 \le \tau(t) \le r$

Into the formula (5.42) instead of ξ one substitutes the new variable $-\xi + \tau(t)$. After calculations the formula is attained

$$\left(\frac{d\tau(t)}{dt} - 1\right)\frac{d\beta(\xi)}{d\xi} - \frac{d\beta^T(-\xi+\tau(t))}{d\xi}C = \left(\frac{d\tau(t)}{dt} - 1\right)^{-1}B^T(-\xi+\tau(t))B - A^T\beta(\xi) \quad (5.43)$$

In this way one obtained the set of differential equations

$$\begin{cases} C^T\frac{d\beta(\xi)}{d\xi} - \left(\frac{d\tau(t)}{dt} - 1\right)\frac{d\beta^T(-\xi+\tau(t))}{d\xi} = \left(\frac{d\tau(t)}{dt} - 1\right)^{-1}B^T\beta(\xi) - \beta^T(-\xi+\tau(t))A \\ \left(\frac{d\tau(t)}{dt} - 1\right)\frac{d\beta(\xi)}{d\xi} - \frac{d\beta^T(-\xi+\tau(t))}{d\xi}C = \left(\frac{d\tau(t)}{dt} - 1\right)^{-1}B^T(-\xi+\tau(t))B - A^T\beta(\xi) \end{cases} \quad (5.44)$$

for $t \ge t_0$, $\xi \in [0, \tau(t)]$ where $0 \le \tau(t) \le r$ with the initial conditions $\beta(0)$ and $\beta(\tau(t))$.
One can reshape the set of equations (5.44) to the form

$$\begin{cases} C^T\frac{d\beta(\xi)}{d\xi}C - \left(\frac{d\tau(t)}{dt} - 1\right)^2\frac{d\beta(\xi)}{d\xi} = \left(\frac{d\tau(t)}{dt} - 1\right)A^T\beta(\xi) + \\ \quad + \left(\frac{d\tau(t)}{dt} - 1\right)^{-1}B^T\beta(\xi)C - \beta^T(-\xi+\tau(t))(AC+B) \\ C^T\frac{d\beta(-\xi+\tau(t))}{d\xi}C - \left(\frac{d\tau(t)}{dt} - 1\right)^2\frac{d\beta(-\xi+\tau(t))}{d\xi} = \beta^T(\xi)(AC+B) + \\ \quad - \left(\frac{d\tau(t)}{dt} - 1\right)A^T\beta(-\xi+\tau(t)) - \left(\frac{d\tau(t)}{dt} - 1\right)^{-1}B^T\beta(-\xi+\tau(t))C \end{cases} \quad (5.45)$$

for $t \ge t_0$, $\xi \in [0, \tau(t)]$ where $0 \le \tau(t) \le r$ with the initial conditions $\beta(0)$ and $\beta(\tau(t))$
There holds the relationship between $\beta(\xi)$ and $\beta(-\xi+\tau(t))$

$$\beta(\xi)\big|_{\xi=\frac{\tau(t)}{2}} = \beta(-\xi+\tau(t))\big|_{\xi=\frac{\tau(t)}{2}} \qquad (5.46)$$

The derivative of the equation (5.24) with respect to t is calculated

$$\left(\frac{d\alpha(t)}{dt} + \frac{d\alpha^T(t)}{dt}\right)(AC+B) + \frac{d\beta(\tau(t))}{dt}C + \frac{d\beta(0)}{dt}\left(\frac{d\tau(t)}{dt} - 1\right) + \frac{d^2\tau(t)}{dt^2}\beta(0) = 0 \qquad (5.47)$$

where

$$\frac{d\beta(0)}{dt} = \frac{d\beta(\xi)}{d\xi}\frac{d\tau(t)}{dt}\Big|_{\xi=0} \tag{5.48}$$

$$\frac{d\beta(\tau(t))}{dt} = \frac{d\beta(\xi)}{d\xi}\frac{d\tau(t)}{dt}\Big|_{\xi=\tau(t)} \tag{5.49}$$

The equation (5.45) implies

$$C^T\frac{d\beta(0)}{dt}C - \left(\frac{d\tau(t)}{dt}-1\right)^2\frac{d\beta(0)}{dt} = \frac{d\tau(t)}{dt}\left(\frac{d\tau(t)}{dt}-1\right)A^T\beta(0) +$$

$$+ \frac{d\tau(t)}{dt}\left(\frac{d\tau(t)}{dt}-1\right)^{-1}B^T\beta(0)C - \frac{d\tau(t)}{dt}\beta^T(\tau(t))(AC+B) \tag{5.50}$$

$$C^T\frac{d\beta(\tau(t))}{dt}C - \left(\frac{d\tau(t)}{dt}-1\right)^2\frac{d\beta(\tau(t))}{dt} = \frac{d\tau(t)}{dt}\beta^T(0)(AC+B) +$$

$$- \frac{d\tau(t)}{dt}\left(\frac{d\tau(t)}{dt}-1\right)A^T\beta(\tau(t)) - \frac{d\tau(t)}{dt}\left(\frac{d\tau(t)}{dt}-1\right)^{-1}B^T\beta(\tau(t))C \tag{5.51}$$

From equation (5.23) one obtains

$$\frac{d\alpha(t)}{dt} = -A^T\alpha(t) - \alpha(t)A - \beta(\tau(t)) - G \tag{5.52}$$

One puts the term (5.52) into the equation (5.47). After calculations one gets

$$\left[A^T\left(\alpha(t)+\alpha^T(t)\right) + \left(\alpha(t)+\alpha^T(t)\right)A\right](AC+B) + \left(\beta(\tau(t))+\beta^T(\tau(t))\right)(AC+B) +$$

$$- \frac{d^2\tau(t)}{dt^2}\beta(0) - \frac{d\beta(\tau(t))}{dt}C - \frac{d\beta(0)}{dt}\left(\frac{d\tau(t)}{dt}-1\right) = -\left(G+G^T\right)(AC+B) \tag{5.53}$$

The matrix $\alpha(t)$, the initial conditions of the system (5.45) and $\frac{d\beta(0)}{dt}$, $\frac{d\beta(\tau(t))}{dt}$ are obtained by solving the set of algebraic equations (5.53), (5.24), (5.50), (5.57) and (5.46). That set of the equations is written below

$$\left[A^T\left(\alpha(t)+\alpha^T(t)\right) + \left(\alpha(t)+\alpha^T(t)\right)A\right](AC+B) + \left(\beta(\tau(t))+\beta^T(\tau(t))\right)(AC+B) +$$

$$- \frac{d^2\tau(t)}{dt^2}\beta(0) - \frac{d\beta(\tau(t))}{dt}C - \frac{d\beta(0)}{dt}\left(\frac{d\tau(t)}{dt}-1\right) = -\left(G+G^T\right)(AC+B) \tag{5.54}$$

$$\left(\alpha(t)+\alpha^T(t)\right)(AC+B) + \beta(\tau(t))C + \beta(0)\left(\frac{d\tau(t)}{dt}-1\right) = 0 \tag{5.55}$$

$$C^T\frac{d\beta(0)}{dt}C - \left(\frac{d\tau(t)}{dt}-1\right)^2\frac{d\beta(0)}{dt} = \frac{d\tau(t)}{dt}\left(\frac{d\tau(t)}{dt}-1\right)A^T\beta(0) +$$

$$+\frac{d\tau(t)}{dt}\left(\frac{d\tau(t)}{dt}-1\right)^{-1}B^T\beta(0)C-\frac{d\tau(t)}{dt}\beta^T(\tau(t))(AC+B) \tag{5.56}$$

$$C^T\frac{d\beta(\tau(t))}{dt}C-\left(\frac{d\tau(t)}{dt}-1\right)^2\frac{d\beta(\tau(t))}{dt}=\frac{d\tau(t)}{dt}\beta^T(0)(AC+B)+$$

$$-\frac{d\tau(t)}{dt}\left(\frac{d\tau(t)}{dt}-1\right)A^T\beta(\tau(t))-\frac{d\tau(t)}{dt}\left(\frac{d\tau(t)}{dt}-1\right)^{-1}B^T\beta(\tau(t))C \tag{5.57}$$

$$\beta(\xi)\,|_{\xi=\frac{\tau(t)}{2}}=\beta(-\xi+\tau(t))\,|_{\xi=\frac{\tau(t)}{2}} \tag{5.58}$$

Having the solution of the set of differential equations (5.45) and taking into account the formulas (5.28), (5.38) and (5.41) one can get the matrices

$$\beta(\theta+\tau(t))=\beta(\xi)\,|_{\xi=\theta+\tau(t)} \tag{5.59}$$

$$\delta(\theta+\tau(t),\sigma+\tau(t))=f(\sigma-\theta) \tag{5.60}$$

where

$$f(\rho)=-\left(\frac{d\tau(t)}{dt}-1\right)\frac{d\beta^T(\rho+\tau(t))}{d\rho}-\beta^T(\rho+\tau(t))A \tag{5.61}$$

for $t\geq t_0$; $\theta\in[-\tau(t),0]$; $\sigma\in[\theta,0]$ where $0\leq\tau(t)\leq r$.

In this way one obtained all coefficients of the functional (5.19). This coefficients depend on the matrices A, B and C of the system (5.9). The time derivative of the functional (5.19) is negative definite. When the matrices $\alpha(t)$, $\beta(\theta+\tau(t))$ and $\delta(\theta+\tau(t),\sigma+\tau(t))$ for $t\geq t_0$, $\theta\in[-\tau(t),0],\sigma\in[\theta,0]$ are positive definite the functional (5.19) becomes the Lyapunov functional.

5.4   An example. An inertial system with a delay and a PD controller.

Let us consider a first order inertial system with a delay described by an equation

$$\begin{cases}\frac{dx(t)}{dt}=-\frac{q}{T}x(t)+\frac{k_0}{T}u(t-\tau(t))\\x(t_0)=x_o\\x(t_0+\theta)=0\\u(t)=-kx(t)-T_d\frac{dx(t)}{dt}\end{cases} \tag{5.62}$$

$t\geq t_0$, $x(t)\in\mathbb{R}$, $\theta\in[-r,0)$, k, k_0, T, T_d, q, $x_0\in\mathbb{R}$, $\tau(t)$ is a time-varying delay satisfying the condition $0\leq\tau(t)\leq r$, $\frac{d\tau(t)}{dt}\neq 1$ where r is positive constant. The parameter k_0 is a gain of a plant, k is a proportional gain, T_d is a derivative gain, T is a system time constant, x_0 is an initial state of a system. In the case $q=1$ an equation (5.62) describes a static object and in the case $q=0$ an equation (5.62) describes an astatic object.

One can reshape an equation (5.62) to a form

$$\begin{cases} \frac{dx(t)}{dt} + \frac{k_0 T_d}{T} \frac{dx(t-\tau(t))}{dt} = -\frac{q}{T}x(t) - \frac{k_0 k}{T}x(t-\tau(t)) \\ x(t_0) = x_o \\ x(\theta) = 0 \end{cases} \qquad (5.63)$$

for $t \geq t_0$ and $\theta \in [-r,0)$.

It is assumed that the element $\frac{k_0 T_d}{T}$ satisfies a condition (3.15), whose takes a form

$$\left| \frac{k_0 T_d}{T} \right| < 1 \qquad (5.64)$$

A new function y is introduced and defined by term

$$y(t) = x(t) - Cx(t-\tau(t)) \qquad (5.65)$$

for $t \geq t_0$.

One can reshape the equation (5.63) to the form

$$\begin{cases} \frac{dy(t)}{dt} = -\frac{q}{T}y(t) + \left(\frac{q k_0 T_d}{T^2} - \frac{k_0 k}{T} \right) x(t-\tau(t)) \\ y(t) = x(t) + \frac{k_0 T_d}{T}x(t-\tau(t)) \\ y(t_0) = x_0 \\ x(t_0 + \theta) = 0 \end{cases} \qquad (5.66)$$

The Lyapunov functional is given by a formula

$$V(y(t),x_t,t) = \alpha(t)y^2(t) + \int_{-\tau(t)}^{0} \beta(\theta+\tau(t))y(t)x_t(\theta)\,d\theta +$$

$$+ \int_{-\tau(t)}^{0} \int_{\theta}^{0} \delta(\theta+\tau(t),\sigma+\tau(t))x_t(\theta)x_t(\sigma)\,d\sigma d\theta \qquad (5.67)$$

where

$$x_t(\theta) = x(t+\theta)$$

for $\theta \in [-r,0)$, $x_t \in W^{1,2}([-r,0),\mathbb{R})$

The coefficients of the functional (5.67) will be obtained.

The equation (5.45) takes the form

$$\begin{bmatrix} \frac{d\beta(\xi)}{d\xi} \\ \frac{d\beta(-\xi+\tau(t))}{d\xi} \end{bmatrix} = \begin{bmatrix} p_1 & -p_2 \\ p_2 & -p_1 \end{bmatrix} \begin{bmatrix} \beta(\xi) \\ \beta(-\xi+\tau(t)) \end{bmatrix} \qquad (5.68)$$

for $t \geq t_0$, $\xi \in [0, \tau(t)]$, where $0 \leq \tau(t) \leq r$

$$p_1 = \frac{-\frac{q}{T}\left(\frac{d\tau(t)}{dt} - 1\right) + \frac{k_0^2 k T_d}{T^2\left(\frac{d\tau(t)}{dt} - 1\right)}}{\frac{k_0^2 T_d^2}{T^2} - \left(\frac{d\tau(t)}{dt} - 1\right)^2} \tag{5.69}$$

$$p_2 = \frac{\frac{q k_0 T_d}{T^2} - \frac{k_0 k}{T}}{\frac{k_0^2 T_d^2}{T^2} - \left(\frac{d\tau(t)}{dt} - 1\right)^2} \tag{5.70}$$

The fundamental matrix of the differential equation (5.68) is given by formula

$$R(\xi) = \begin{bmatrix} ch\lambda\xi + \frac{p_1}{\lambda}sh\lambda\xi & -\frac{p_2}{\lambda}sh\lambda\xi \\ \frac{p_2}{\lambda}sh\lambda\xi & ch\lambda\xi - \frac{p_1}{\lambda}sh\lambda\xi \end{bmatrix} \tag{5.71}$$

where

$$\lambda = \frac{\sqrt{\frac{k_0^2 k^2 - q^2\left(\frac{d\tau(t)}{dt} - 1\right)^2}{\frac{k_0^2 T_d^2}{T^2} - \left(\frac{d\tau(t)}{dt} - 1\right)^2}}}{T\left(\frac{d\tau(t)}{dt} - 1\right)} \tag{5.72}$$

Hence

$$\begin{bmatrix} \beta(\xi) \\ \beta(-\xi + \tau(t)) \end{bmatrix} = R(\xi) \begin{bmatrix} \beta(0) \\ \beta(\tau(t)) \end{bmatrix} \tag{5.73}$$

for $t \geq t_0$, $\xi \in [0, \tau(t)]$ where $0 \leq \tau(t) \leq r$.

One needs the initial conditions of the set of differential equations (5.68) to obtain

$$\beta(\theta + \tau(t)) = \beta(\xi)|_{\xi = \theta + \tau(t)} \tag{5.74}$$

$$\delta(\theta + \tau(t), \sigma + \tau(t)) = f(\sigma - \theta) \tag{5.75}$$

$$f(\rho) = -\left(\frac{d\tau(t)}{dt} - 1\right)\frac{d\beta(\rho + \tau(t))}{d\rho} - a\beta(\rho + \tau(t)) \tag{5.76}$$

for $t \geq t_0$, $\theta \in [-\tau(t), 0]$, $\sigma \in [\theta, 0]$ where $0 \leq \tau(t) \leq r$.

The initial conditions of the differential equation (5.68) and the coefficient $\alpha(t)$ are attained by solving of a set of equations (5.54) to (5.58) which take the form as below

$$4a\left(\frac{qk_0 T_d}{T^2} - \frac{k_0 k}{T}\right)\alpha(t) + \left(\frac{k_0 T_d}{T}p_2\frac{d\tau(t)}{dt} - \frac{d^2\tau(t)}{dt^2} - p_1\frac{d\tau(t)}{dt}\left(\frac{d\tau(t)}{dt} - 1\right)\right)\beta(0) +$$

$$+ \left(2\left(\frac{qk_0 T_d}{T^2} - \frac{k_0 k}{T}\right) - \frac{k_0 T_d}{T}p_1\frac{d\tau(t)}{dt} + p_2\frac{d\tau(t)}{dt}\left(\frac{d\tau(t)}{dt} - 1\right)\right)\beta(\tau(t)) =$$

$$= -2w \left(\frac{qk_0 T_d}{T^2} - \frac{k_0 k}{T} \right) \tag{5.77}$$

$$2 \left(\frac{qk_0 T_d}{T^2} - \frac{k_0 k}{T} \right) \alpha(t) + \left(\frac{d\tau(t)}{dt} - 1 \right) \beta(0) - \frac{k_0 T_d}{T} \beta(\tau(t)) = 0 \tag{5.78}$$

$$\left(ch\frac{\lambda \tau(t)}{2} + \frac{p_1 - p_2}{\lambda} sh\frac{\lambda \tau(t)}{2} \right) \beta(0) + \left(\frac{p_1 - p_2}{\lambda} sh\frac{\lambda \tau(t)}{2} - ch\frac{\lambda \tau(t)}{2} \right) \beta(\tau(t)) = 0 \tag{5.79}$$

Bibliography

[1] J. Duda: Parametric optimization of neutral linear system with respect to the general quadratic performance index. Archiwum Automatyki i Telemechaniki, 1988, 33, (3), pp. 448-456

[2] J. Duda: Lyapunov functional for a linear system with two delays both retarded and neutral type. Archives of Control Sciences, 2010, 20, (1), pp. 89-98.

[3] J. Duda: Lyapunov functional for a system with k-non-commensurate neutral time delays. Control and Cybernetics, 2010, 39, (4), pp. 1173-1184.

[4] J. Duda:Parametric optimization of neutral linear system with two delays with P-controller. Archives of Control Sciences, 2011, 21, (4), pp. 363–372

[5] J. Duda: Lyapunov functional for a linear system with both lumped and distributed delay, Control and Cybernetics, 2011, 40, (1), pp. 73-90

[6] J. Duda: Lyapunov functional for a system with a time-varying delay, International Journal of Applied Mathematics and Computer Science, 2012, 22, (2), pp. 327-337

[7] J. Duda:Parametric optimization of a neutral system with two delays and PD-controller, Archives of Control Sciences, 2013, 23, (2), pp. 131–143

[8] J. Duda: A Lyapunov functional for a neutral system with a time-varying delay, Bulletin of The Polish Academy of Sciences Technical Sciences, 2013, 61, (4), pp. 911-918

[9] H. Górecki, S. Fuksa, P. Grabowski, A. Korytowski: Analysis and Synthesis of Time Delay Systems, John Wiley & Sons, Chichester, New York, Brisbane, Toronto, Singapore, 1989

[10] J. Hale, S. Verduyn Lunel: Introduction to Functional Differential Equations. New York, Springer, 1993

[11]   Yu. M. Repin: Quadratic Lyapunov functionals for systems with delay. Prikl. Mat. Mekh. 29(1965), 564-566